Hypercity

Cities are sites of multiple meanings and symbols, ranging from statues and street names to festivals and architecture. Sometimes the symbolic side of urbanism is so strong that it outshines reality – then we speak of *hypercity*. Urban symbolic ecology and hypercity studies are relatively new fields that deal with the production, distribution and consumption of symbols and meanings in urban space, timely concerns in an era of increasing globalization and competition between mega-urban regions. This volume presents a detailed introduction to the new fields, followed by case studies of the cultural layer of symbolism in Brussels (Belgium), Cape Town (South Africa), Cuenca (Ecuador), Delft (The Netherlands), Kingston (Jamaica), Ljubljana (Slovenia), Paris (France) and cities in Italy and Indonesia. It amply demonstrates that the time has come for urban symbolic ecology and hypercity studies to be included in regular urban studies training in the fields of anthropology, sociology, social geography and architecture.

The Editors
Peter J.M. Nas is an urban anthropologist and Professor in the Socio-cultural Aspects of the Built Environment in Indonesia at Leiden University, the Netherlands. **Annemarie Samuels** is an MA student in Cultural Anthropology and Development Sociology at Leiden University, The Netherlands.

www.keganpaul.com

Studies in Anthropology, Economy and Society

~

Frontispiece: The patron saint of San Gimignano (Italy) with the city crowned by feudal towers on his lap; late fourteenth–century altarpiece by Taddeo di Bartolo (Source: Enzo Rafa, n.d., *San Gimignano by the Beuatiful Towers: Guide-Book Art and History*, p. 1. San Gimignano: Brunello Granelli's Edition).

Hypercity

The Symbolic Side of Urbanism

Edited by

Peter J.M. Nas
and
Annemarie Samuels

KEGAN PAUL
London • Bahrain • New York

First published in 2006 by
Kegan Paul Limited
UK: P.O. Box 256, London WC1B 3SW, England
Tel: 020 7580 5511 Fax: 020 7436 0899
E-mail: books@keganpaul.com
Internet: http://www.keganpaul.com
USA: 61 West 62nd Street, New York, NY 10023
Tel: (212) 459 0600 Fax: (212) 459 3678
Internet: http://www.columbia.edu/cu/cup
BAHRAIN: bahrain@keganpaul.com

Distributed by:
Marston Book Services Ltd.
160 Milton Park
Abingdon
Oxfordshire OX14 4SD
United Kingdom
Tel: (01235) 465500 Fax: (01235) 465555
E-mail: direct.orders@marston.co.uk

Columbia University Press
61 West 62nd Street, New York, NY 10023
Tel: (212) 459 0600 Fax: (212) 459 3678
Internet: http://www.columbia.edu/cu/cup

ISBN-10: 0-7103-1279-2
ISBN-13: 978-0-7103-1279-2

British Library Cataloguing in Publication Data

Hypercity : the symbolic side of urbanism. – (Studies in anthropology, economy and society)
1. Sociology, Urban 2. Social ecology 3. Symbolism in city planning 4. Symbolism in architecture
I. Nas, P. II. Samuels, Annemarie
307.7'6

ISBN-10: 0710312792

Contents

Contents

Contributors

Roelf ter Berg finished his BA and MA studies at the Department of Cultural Anthropology and Development Sociology, Leiden University. During his MA study he conducted field work in Cape Town, South Africa.

Alexandra Bitusikova is a social anthropologist (PhD Comenius University in Bratislava, Slovakia). Since 1991 she has been working as a senior research fellow and director (1993–2000) at the Institute of Social and Cultural Studies of Matej Bel University in Banska Bystrica, Slovakia. She has lectured at many conferences throughout Europe. She is author of *Urban Anthropology: Tendencies and Perspectives* (in Slovak), a number of publications focusing on urban anthropology, post-socialist social and cultural change in Slovakia, identities, minorities and gender. In 2001–2002 she worked for the European Commission, DG Research, Ethics of Science and Research Unit. Now based in Brussels, she works as a senior project manager at the European University Association and continues to work for Matej Bel University on several EU research projects.

Freek Colombijn (1961) is an anthropologist and historian (PhD Leiden University, 1994). He is co-editor (with J. Thomas Lindblad) of *Roots of Violence in Indonesia: Contemporary Violence in Historical Perspective* (Leiden, KITLV Press, 2002), and (with Aygen Erdentug) of *Urban Ethnic Encounters: The Spatial Consequences* (London, Routledge, 2002). His research interests include urban development, environmental history, and the politics of road construction. He is currently assistant professor at the Department of Social and Cultural Anthropology of the Vrije Universiteit in Amsterdam, the Netherlands.

Rivke Jaffe is a junior researcher at Leiden University, where she is currently completing her PhD thesis on urban environmental problems in the Caribbean, following extensive fieldwork in Jamaica and Curaçao. She is scheduled to begin a post-doc on class and ethnicity in Surinam at the KITLV (Royal Netherlands Institute of Southeast Asian and Caribbean Studies). She is editor of *The Caribbean City* (Kingston, Ian Randle, forthcoming) and co-editor (with Peter J.M. Nas and Gerard Persoon) of *Framing Indonesian Realities* (Leiden, KITLV Press, 2003). Her main research interests are urban anthropology and sociology, Caribbean studies and environmental studies.

Božidar Jezernik is professor of anthropology at the University of Ljubljana and dean of the Faculty of Arts. He has published books on Italian, German, and Yugoslav concentration camps: *Struggle for Survival* (in Slovenian, Ljubljana, 1983, English version, Ljubljana, 1999), *Sex and Sexuality in Extremis* (in Slovenian, Ljubljana, 1993), *Non Cogito Ergo Sum* (in Slovenian, Ljubljana, 1994). His book on Balkan travel reports was published in Slovenian in 1998 and translated into Bosnian in 2000. Its expanded and revised English version was published under the title *Wild Europe: The Balkans in the Gaze of Western Travellers* (London, Saqi Books, 2004).

Christien Klaufus (1967) is a cultural anthropologist and an architect. Her research field is urban anthropology and she is especially interested in the socio-cultural and physical construction of space, place and architecture. She is the author of 'Dwelling as Representation: Values of Architecture in an Ecuadorian Squatter Settlement' (*Journal of Housing and the Built Environment* 15, 2000) and 'Globalization in residential architecture in Cuenca, Ecuador: Social and cultural diversification of architects and their clients' (*Environment and Planning: Society and Space*, forthcoming). She is currently completing her dissertation on perceptions and practices regarding popular architecture in Ecuador.

Peter J.M. Nas (1944) is professor at Leiden University, employed at the departments of Cultural Anthropology and Development Sociology, and Languages and Cultures of Southeast Asia and Oceania. He teaches urban anthropology and development sociology, and has conducted extensive fieldwork in Indonesia. He is editor of several works on the Indonesian built environment, the most recent being *Jakarta-Batavia: Socio-cultural Essays* (Leiden, KITLV Press, 2000, with Kees Grijns), *The Indonesian Town Revisited* (Münster, Lit Verlag, 2002), *Framing Indonesian Realities* (Leiden, KITLV Press, 2003, with Gerard Persoon and Rivke Jaffe) and *Indonesian Houses* (Leiden, KITLV Press, 2003, with Reimar Schefold and Gaudenz Domenig). Peter Nas is secretary-general of the International Union of Anthropological and Ethnological

Sciences (IUAES) and chairman of the Association KITLV (Royal Netherlands Institute of Southeast Asian and Caribbean Studies).

Dirk B.J. Noordman (1944) is an economist and philosopher. He is currently director of Culture and Tourism, Risbo Contractresearch BV, at the Erasmus University, Rotterdam, and associate professor in arts management at the Faculty of History and Cultural Studies. He teaches arts management and city marketing at the Erasmus University in Rotterdam, the University of Utrecht and the University of Maastricht. His publications include the books *Kunstmanagement: Een Inleiding* (Art Management: An Introduction, Elsevier/Reed, third edition, 2003), *Museummarketing* (Museum Marketing, Elsevier, Den Haag, 2000) and *Cultuur in de Citymarketing* (Culture in City Marketing, Elsevier/Reed Business Publications, Den Haag, 2004).

Leeke Reinders (1971) works at the OTB Research Institute for Housing, Urban and Mobility Studies, University of Technology in Delft. His publication topics and research interests include the social production and construction of place, processes of urban regeneration, youth culture in the Parisian banlieue and urban developments in Southwest China. He is currently doing an ethnographic PhD study of identity strategies and identification in a post-war neighbourhood in the Netherlands.

Annemarie Samuels is an MA student at the Department of Cultural Anthropology and Development Sociology, Leiden University.

Margot Nicoline te Velde (1971) finished her studies in Sociology of Non-Western Societies at Leiden University in 1998. She wrote her MA thesis on 'Urban Symbolism in Cape Town, South Africa' following fieldwork in 1996. She has been living and working as an executive personal assistant in Cape Town since 2002.

Gaby van der Zee (1980) is studying Languages and Cultures of Latin America at Leiden University. She has conducted fieldwork in Buenos Aires, Argentina, on the subject of ethnicity amongst Italo-Argentines, combining her twin passions for Latin America and Italy. After finishing her studies, she hopes to develop a career in journalism.

Acknowledgements

This book *Hypercity: The Symbolic Side of Urbanism* is based on a number of papers presented at the international workshop 'Urban Symbolism' held in Leiden, 16–18 June 2004. The volume does not contain all the papers presented and we want to express our gratitude to all the participants in the workshop for their work and contributions to the discussions. Rivke Jaffe, MA, of the Research School CNWS, Leiden, played a prominent role in the concluding debate of the workshop. We are grateful to her for suggesting the title of this volume: Hypercity.

We have tried to reach all persons and institutions with copyrights on the illustrations depicted in this book. Those persons who in spite of our efforts still want to claim copyrights are requested please to contact the editors.

Peter J.M. Nas and Annemarie Samuels
Leiden University
1 October 2005

Chapter 1

URBAN SYMBOLIC ECOLOGY AND THE HYPERCITY
State of the art and challenges for the future

Peter J.M. Nas, Rivke Jaffe and Annemarie Samuels

Introduction

While the study of urban symbolism is deeply rooted in history, it only gained momentum in the 1990s, during which period a series of urban symbolism case studies were completed, resulting in a variety of noteworthy articles and monographs.[1] These studies approached each city in an open way and attempted to unveil basic themes underlying its symbolic configuration. The multitude of single-city studies stimulated the development of a large number of concepts. These concepts enabled easy and interesting fieldwork in any city deemed worthwhile, but also produced a certain level of theoretical complacency. While additional case studies do lead to the creation of new concepts, more fundamental reflection is necessary to create further challenges. One way to reach a new level of inspiration is by considering urban symbolism in relation to the study of symbolism in general, looking to the whole field and its basis in semiotics.

This introduction examines the state of the art of urban symbolism, while proposing new directions and challenges for the future of the field. Using two related approaches, urban symbolic ecology and the hypercity, we give an overview of theoretical concepts and their historical and empirical basis. Urban symbolic ecology has its basis in the study of urban material culture, while the hypercity is a new encompassing theoretical framework which regards urban symbolism starting from ideas, images and representations of the city. Both complementary approaches will be elaborated below and

---

[1] See for example Nas 1990, 1993, 1998; Colombijn 1994; Nas and Sluis 2002; Nas and Persoon 2003; Nas and Van Bakel 2003.

linked to the contributions in this volume and to previous research in the broader field of urban symbolism. In doing this, we discuss a number of possible new approaches, including research based on new concepts or specific theoretical key questions, and related to types of cities or crucial events.

Urban symbolic ecology

The material approach to urban symbolism, also known as urban symbolic ecology, takes its point of departure in material culture, starting with a physical entity to elucidate basic value patterns encountered in the city. Several contributions in this volume emphasize the material approach of urban symbolic ecology, which is defined as the social production and consumption along with the spatial distribution of symbols and rituals. In order to cover the whole configuration of urban perception, three concepts are proposed by Kevin Lynch (1975): identity (his term for architectural individuality), structure and meaning. In contrast to Lynch's narrow focus on identity and structure, which disregards meaning because of its individual character, studies in urban symbolic ecology show the importance of meaning for urban perception, orientation and symbolism (Nas and Sluis 2002). Furthermore, the concepts of identity (perception of the self) and image (perception by others); urban memory (collective, historic, projective or cultural); and collective memory in relation to places of generation, trauma and memory[2] are relevant to this symbolic approach of the city. In the context of the distribution of symbols, their different locations in the urban centre or periphery and the total spatial configuration are important.

In the study of urban symbolism through 'grounded' research, many other concepts have been developed to describe symbolic phenomena, systems and patterns of change, with the basic aim of unveiling a fundamental symbolic theme of the urban community under consideration, which might be nation building, ethnic tension and struggle, cosmology, or modernization.

One category[3] of concepts aims to describe the character of symbols. Symbols can be either private or public. They can be sacral or profane: a distinction is made between religious and secular symbols. Symbols can also be multivocal, as they denote different things to different (groups of) people. Symbolism can function at various levels simultaneously: symbols

---

[2] Cf. Pierre Nora's *lieux de mémoire*; Nora 1984-1992.
[3] These categories are not mutually exclusive and other classifications are possible.

are linked to an institutional level of the society (e.g. national or local) or are related to a certain group of people. The meaning of symbols is changeable, for example when power structures or ideologies are transformed. This process is captured by the term symbolic dynamics. When one symbol refers to another (more powerful or widely known) symbol, this is called dependence or referential symbolism. Symbolic saturation describes how symbols become meaningless when people see them too often. When individuals are confronted with a symbol on a daily basis, they may cease to notice it. Another related concept is symbolic force: some symbols may be more powerful and have more impact than others. In addition, symbols can be open or canned:[4] some symbols invite people to react to the symbols or to be actively involved with them, while others render spectators passive.

Another category exists of concepts that relate to types of symbol carriers. In the first place the symbol carrier (or bearer) itself refers to the material symbols, which can be grouped in symbolic categories (for instance, equestrian statues) or broken down into symbolic elements (such as a particular statue) that combine with other elements to make one symbol carrier. These elements can be rooted in a certain symbolic domain, such as the religious, state, economic or educational domain. Nested symbolism indicates that a city has a symbol reinforced through various symbol carriers. For example a city may have statues, monuments, street names, songs, and myths dedicated to one particular leader, a famous artist or a known scientist. In Petersburg the symbolic force of Tsar Peter is strengthened in this way through the city's name, his statue and other elements (such as a poem) referring to this leader who aimed to open up Russia to the West. The variety of types of symbol carriers will be treated more elaborately in the context of the hypercity concept.

Meinig (1979) proposes the concept of symbolic landscapes in contrast to that of concrete landmarks. He analyzes current idealizations of urban landscapes in a general, abstract way through descriptions of the characteristics of the 'main street', the 'suburb' and the 'New England village' as perceived in the United States. These representations are considered in the context of national identity and are related to unambiguous American sentiments and perceptions in this respect. This interesting approach is applied to Tokyo by Cybriwsky (1997), who aims to portray urban historical and present-day landscapes, such as the Tokyo district of Shinjuku, presenting their image and representation. In the case

---

[4] Cf. Scott 1998.

of Indonesian cities, we think this approach could fruitfully be applied to the description of representations of the 'kampong', 'China town', the 'city square' (*alun-alun*), the 'palace' (*kraton*) and the 'mall'.

A next set of concepts is related to power relations and the involvement of different social groups in symbolism. The symbol maker is the producer of a symbol. Symbols can be formal and informal, depending on whether they are official or not. Besides the official meaning of a symbol, unofficial meanings may be attached to it. When symbols are appropriated in order to attack the dominant discourse we can speak of contra-symbolism. A distinction can be made between top-down symbolism (when symbols are imposed from above) and bottom-up symbolism (when symbols have been promoted from the grassroots level). Symbolism is bridled when the authorities limit the symbolic public space of certain groups in society. As a matter of course, symbolism has a function. It can legitimate regimes and certain actions, foster communal or national feelings, and express counter-opinions.

Certain concepts can be used to describe the entire symbolic configuration of a city. Cities may have a historically tiered symbolic system, in which case groups of symbols associated with specific historic periods constitute the urban symbolic system. Symbolic systems can be characterized as either crystallized or emergent. In a crystallized system, the city has one characteristic that defines the entire symbolic configuration, for example in a university city, a harbor city or a garrison city. This pattern can also be emergent, developing towards a crystallized system. Symbolic systems can be uniform or pluriform, depending on whether the system is controlled by one group or rather by several groups of people.

It is clear that the cultures of specific regions, ethnic groups and religions often result in the creation of particular repertoires of symbol bearers. Repertoires vary from city to city, but variations at broader national and cultural levels can be distinguished as well. The constellation of urban markers is influenced by the specific context in which a city functions; whereas many Western cities include a number of statues, Islamic cities tend to embrace other symbol carriers than figurines. Hindu cities have temples and are oriented towards the holy mountain, while Christian cities have a repertoire that is formed by churches and the protector or saint of the town. Italian cities often have tall family towers expressing the medieval culture of the city states, as demonstrated by Gaby van der Zee in this volume. Colonial cities often display a gridiron street pattern. These are just a few examples related to the concept of repertoires of urban symbol bearers, a direction that calls for more elaborate and systematic study.

In the context of urban social change, several development paths in the field of symbolism can be distinguished in which tradition plays a variable role. Unequivocal urban modernization, as exemplified by Singapore, can be represented as follows: A>>>>>B>>>>>C. When urban modernization or westernization aims at more utopian targets, as in Brasília (an 'instant city' with a non-urban starting point) this development path can be represented as: (A>>>>>)B>>>>>C'. In case of social change referring to the traditional situation, as exemplified by some counter-communities in Grozny with the sincere drive to return to the former family group system, this can be represented as: A>>>>>B>>>>>A. A hybrid development path, as is the case in Banda Aceh which aims at 'new' traditions in ethnic architecture, could be represented by: A>>>>>B>>>>>A'. A last hypothetical development path is a pure continuation of the present situation without any reference to the past or the future: A>>>>>B>>>>>B, a possibility which presumably automatically turns into A>>>>>B>>>>>B', as unadulterated continuity without any change is difficult to perceive. This series of development paths should not be considered complete; other variations may be found in further research.

The physical entities, or material culture, from which urban symbolic ecology mostly departs, are the symbol carriers or material signifiers. The papers in this volume by Christien Klaufus, Gaby van der Zee, Božidar Jezernik and Rivke Jaffe each focus on one particular type of material symbol carrier, respectively architecture, the skyline and statues. In her contribution *'Megaproyecto El Barranco' in Cuenca, Ecuador: Architects and their role in local representations* Christien Klaufus shows how architectural representations and cultural symbols are embedded in an Andean upper-class worldview. The decision of the mayor of Cuenca to organize a national design competition to generate ideas for one of the most emblematic areas of Cuenca, 'El Barranco', provoked strong debate among local architects. This national competition was meant to stimulate discussion on the area's future development, but also appeared to challenge existing power relations among local architects and urban planners. Through an analysis of the controversy surrounding the design competition, Christien Klaufus unravels local architects' stances on urban design and on the representation of local cultures and social categories in urban space.

In *The Italian skyline: Family towers as elements in urban symbolism* Gaby van der Zee points out that another urban symbol carrier, the skyline, has always been important in the ability of cities to distinguish themselves.

In medieval Italy, city towers formed one of the main features in the skyline. They belonged to a wealthy family or a group of families and functioned as status symbols. Their immense symbolic force expressed a family's cohesion as well as displaying its wealth and authority. These family towers can be found in various documents, chronicles, art and other illustrations (for example the Patron Saint of San Gimignano holding the city in his hands) and are displayed on coins and royal seals. Since the towers symbolized the power of the nobles, it was in the interest of the urban communes or the city-states and rival noble families to destroy them. Today, tall buildings are still erected as status symbols, exhibits of wealth and power to the world. Though this type of buildings has often been transformed from private into public symbols, they still have a strong overall symbolic force, representing the capitalist world. And today, too, these symbols are being destroyed; the 9/11 attacks on the WTC are the best example.

A third important symbol carrier is the statue. In *Looking back, forward or up? Urban symbolism and the politics of race in Jamaica's Redemption Song* Rivke Jaffe examines one single sculpture. The 'Redemption Song' sculpture in Kingston, Jamaica, was intended as a national symbol commemorating the abolition of slavery. The symbol's significance is related to the construction of national identity, a process that is intimately linked to local power relations and ethnic discourse in post-colonial society. Her study centers on the 2003 debate regarding the sculpture, focusing in particular on its symbolic production (what it was intended to signify and express) and consumption (how its meaning was perceived and constructed by the larger public). The discrepancy between these two processes led to a controversy which must be seen in the light of Jamaica's history of slavery and institutionalized racism resulting in the contemporary politics of race.

While Rivke Jaffe restricts herself to one exemplary item in Kingston, Jamaica, Božidar Jezernik, in *Power of remembrance, supremacy of oblivion: History of the 'National Monuments' in Ljubljana*, discusses the whole corpus of statues in this Slovenian town. Over the centuries, public monuments were erected in Ljubljana in order to create and re-create Slovenian history. On the face of it, these public monuments were meant to commemorate historical personalities and events, but in order to shape history some aspects had to be neglected and as a consequence monuments were destroyed. Public monuments became a tool to reconstruct Slovenian history as a linear and straightforward story connecting the present with the mythical past. In this context, all monuments that were expressions of Austrian and Yugoslav patriotism were removed, and after independence the authorities removed all of Tito's monuments. Nowadays, there are no public monuments that testify to Slovenia's historical links with Austria or Yugoslavia.

Hypercity

The concept of the hypercity, a theoretical lens through which to view the city, takes urban symbolism back to the basics, focusing on semiotics. While urban symbolic ecology focuses on material culture, the hypercity takes its starting point in the intangible semiotic realm. At its most basic level, a symbol consists of two elements, the signifier and the signified. The 'sign' or symbol itself lies in the relationship between the signifier and the signified; one without the other is meaningless. These terms were first developed in linguistics by Ferdinand de Saussure; he saw language as consisting of verbal signs, in which a word ('chair') is the signifier, while the signified is the object or idea the word represents (something you sit on). Hence a linguistic sign links sound images with concepts. Meaning lies in the connection between the two, the process of signification. A particular sound may mean something entirely different to English and non-English speakers and the agreement to use certain words is termed a cultural convention, a system of representation which is culturally constructed. In essence, linguistic signs are completely arbitrary; for this reason de Saussure chose not to use the word 'symbol', feeling that it implied a certain motivation. Signs relate to one another and function within sign-systems which incorporate cultural values and codes. Signification, the process of meaning between the signifier and the signified, is culturally loaded. This means that it relates to dominant ideologies. Language, like all other signs, is not neutral; it reflects and reinforces power differentials. Researchers such as Roland Barthes (2000, 2000a) dissected the way in which mass media disseminate powerful ideologies through their manipulation of signs, signifiers and images.

A keystone of postmodernism is 'hyperreality', a concept expounded most powerfully by Baudrillard (1993, 1994) and Umberto Eco (1998). The proliferation of images in our 'mediatized' daily lives have ceased to be a 'reflection of profound reality' and now have 'no relation to any reality whatsoever', they become pure simulacra (Baudrillard 1994: 6). In a media-saturated world full of simulation, representation and fantasy, images become more real than reality, and signifiers become more real than the signified: hyperreality. For instance, one could look at a real-life sunset, and have the feeling that it has something patently kitsch to it. Another example is when one smells a lilac tree and is struck by the fact that it smells like laundry detergent. The trend of television 'reality shows' also supports the passing of heavily edited, elicited televised images for real.

The concept of 'hypercity' applies the phenomenon of hyperreality to the city: the whole of urban symbols creates a reality that can become more real than the city itself. The city – Paris, Rio de Janeiro or Tokyo – is the signified, while the total idea of or rather the total of ideas of these cities constitute their signifiers: their hypercity. The mundane or gritty reality of the millions of people inhabiting these cities becomes subordinate to Paris as a city of culture and romance, Rio as the urban embodiment of Carnival, samba and beaches, and Tokyo as a global centre of modernity and efficiency. The hypercity encompasses the entirety of urban signifiers or symbol carriers that combine in a dynamic process of signification to represent the city, both to its inhabitants and to the rest of the nation-state, the region and the world.

A distinction can be made between material, iconic, behavioural and discursive signifiers, or symbol carriers. Material signifiers are found in the traditional domain of urban symbolic ecology, described above, and include statues, landmarks, architecture, neighbourhoods and both man-made and natural landscapes. The Berlin Wall, Beijing's Forbidden City and Kuala Lumpur's Petronas Towers are all examples of famous material signifiers. Iconic signifiers consist of people, whether individuals or groups, who have become representative for certain cities. Heroes, saints, royalty and celebrities are all iconic figures that can shape this category. Vatican City is associated with the Pope, and Jesus with Nazareth, while Memphis is personified by Elvis. Liverpool is represented by the Beatles, but other groups, in particular sports teams, have even stronger symbolic power: Rotterdam rallies around its soccer team Feyenoord, Baltimore is the Orioles for baseball fans, and basketball has resulted in the Lakers being synonymous with Los Angeles. Behavioural signifiers include rituals, ceremonies, holidays, festivals, demonstrations and parades. The film festival in Cannes, the annual Bull Run in the Spanish city of Pamplona, or anti-globalist demonstrations in Genoa and Seattle illustrate this. Discursive signifiers are less tangible and include urban images and narratives propagated in movies, logos, paintings, songs, poems and novels as well as maps, newspapers, television shows, travel guides, billboards, street names and urban planning policy. Urban legends and myths also contribute to this discursive category. Think of the multitude of novels set in Paris, all the songs extolling the virtues and vices of New York, and how the emblematic representation of the London Underground reshapes mental maps of that city. Inevitably, New Orleans is associated with jazz, Chicago with the blues, Buenos Aires with tango, and Kingston with reggae. The image of New

York is determined to a large extent by the innumerable reproductions of its skyline. Next to the loss of life, the 9/11 attacks were shattering in a symbolic sense, first as an attack on Western modernity, but also because of the irreversible damage to the quintessential image of New York, its skyline. Each individual city will have a different repertoire of signifiers, dependent on its cultural context.

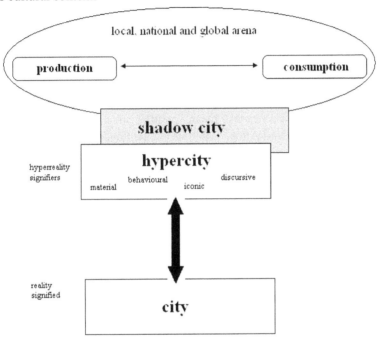

Figure 1 Hypercity and city.

In the hypercity, signification proceeds through the production and consumption of signifiers. Material, iconic, behavioural and discursive signifiers alike are produced by individuals or actors who, consciously or unconsciously, ascribe particular meanings to objects, behaviour and discourse. In turn, other actors are confronted with these objects, behaviour and discourse which to them are imbued with meaning, though this meaning may vary from that intended by the symbol-producers. A symbolically coherent hypercity is one in which production and consumption merge. A divergence of these twin processes, resulting in an incoherent hypercity, is more likely in ethnically, economically or politically fragmented cities. On a hypothetical scale of zero to ten, a ten would be achieved in a city where signs produced are identical to those consumed, and each signifier has one, unanimously accepted meaning. Zero would be a city in which no individual agrees with another on what ideas and images represent the city, and where each individual attributes a different meaning to urban signifiers, shared meaning being absent. Naturally, neither extreme is likely in the 'real world'.

Like language, the meaning of the city and the process of signification are not neutral; rather, the hypercity is contested in a continuous process of symbolic production and consumption in which those with power are able to leave a larger mark. Returning to linguistic analogies, this is illustrated in an encounter Alice has with Humpty Dumpty in Lewis Carroll's *Alice in Wonderland*:

> 'When I use a word,' Humpty Dumpty said in a rather a scornful tone, 'it means just what I choose it to mean – neither more nor less.'
> 'The question is,' said Alice, 'whether you *can* make words mean different things.'
> 'The question is,' said Humpty Dumpty, 'which is to be master – that's all.'

One *can* make cities mean different things, shaping the hypercity, most effectively if one achieves convergence of symbolic production and consumption. These symbolic processes are effected by a variety of stakeholders in local, national and global arenas, often simultaneously. At the local level, such a process might involve a multi-ethnic city in which different ethnic groups attach wildly varying meanings to symbols produced by a largely mono-ethnic municipal government. In the global arena, orientalism and its twin, occidentalism, influence urban image and identity. Occidentalist hardliners view Western cities as concentrations of modernity, decadence and sin. The twentieth century saw Shanghai and Marrakech function as prototypes of exotic Eastern cities. Istanbul, the former Constantinople, is a previously archetypical Eastern city that has come to promote the image of a bridge between East and West. The symbolic arena is most obvious in situations of changing power relations, specifically at the national level.

In *Sign of the times: Symbolic change around Indonesian independence*, Freek Colombijn gives a comparative analysis of symbolism and power relations in Indonesian cities by studying the political process that culminated in the decolonization of Indonesia. The Netherlands East Indies, a Dutch colony, were occupied by the Japanese during the Second World War. Shortly after the Japanese surrender to Allied forces, nationalist forces led by Sukarno and Mohammed Atta, proclaimed the independent Republic of Indonesia, a decision contested militarily by the Dutch until 1949. Focusing on three types of symbol carriers – statues, ceremonies and street names – Colombijn compares symbolic change around independence in different Indonesian cities in an attempt to explain symbolic variation between cities. In doing this, he indicates the variation of meanings attributed to symbol carriers under Dutch, Japanese and Indonesian rule, as well as examining the processes by which new symbols were created. The three types of symbol carriers fared differently: the symbolic consumption of statues tended to be sufficiently pluriform for them to survive political change. In contrast, street names, hegemonic because only the state can officially determine them, were amended rapidly to reflect nationalist priorities. However, this change in symbolic production was undermined by urban consumers who continue to use older street names. Ceremonies including parades and the hoisting of flags display more abrupt and irreversible instances of symbolic change. Comparative analysis reveals that political turns in Indonesia were reflected in changes in different types of symbol carriers, and that these shifts were connected to the different meaning ascribed to various types of symbol carriers by Dutchmen, Japanese and Indonesians.

Interpretation and strategic use of historical events and eras are also powerful tools in the symbolic arena. Urban history can present validations for present and future policies or collective action. In one of this volume's two studies of Cape Town, Peter J.M. Nas, Margot te Velde and Annemarie Samuels explore the divide between natural and man-made urban symbol carriers. They approach the latter through studying material signifiers that refer to four historical periods, respectively the Dutch colonial period, British rule, the Apartheid era and the current period of transition and modernization. The city's man-made urban symbolism reflects social change in its layering, as the various population groups in power during the different periods leave their consecutive marks on the urban landscape. The authors find that the symbolism of these different periods is very clear and crystallized, with the exception of the post-Apartheid period in which they observe a 'symbolic void'. As witnessed by many abortive plans and proposals for new urban symbols, both the state and the city's inhabitants appear to be unprepared to accept a unified symbolic order as yet. With the exception of the symbolic human figure of Nelson Mandela, there are no symbols that are acceptable to all of Cape Town's population groups. However, Cape Town is graced with a number of spectacular natural landmarks, most famously the Devil's Peak, Table Mountain, Lion's Head and Table Bay. The natural environment is imbued with various meanings, related to different values and functions; it clearly constitutes an important part of Cape Town's identity. Natural landmarks display a strong dominance as spontaneous symbols, rather than those that are created or consciously produced. They function as continuous, neutral symbolic features and are accepted by all groups in urban society. In the current phase of transition these natural monuments can substitute formally produced symbols in promoting an inclusive urban identity.

The hypercity is democratic in that everyone is a consumer, and everyone has the potential to be a producer. But not all voices are heard with equal clarity. In the power-laden arena of the hypercity, those actors with more power and access to symbol carriers are obviously more influential in the production of symbols. On the one hand, these producers belong to more traditional power-holders: the state, the media, urban planners, and architects. On the other hand, artists, a group that generally holds a negligible amount of economic power, play a major role in the production of the hypercity as mediators between reality and illusion. Movie directors, rap artists, photographers and poets shape the way a city is perceived by its inhabitants and by the outside world. Jay McInerny's novel *Bright Lights, Big City* eternalized New York in the 1980s, as Fellini's *La Dolce Vita* did mid-twentieth century Rome. The 1960s and 1970s saw songs by a variety of artists put San Francisco on the map as hippy capital of the world. The recent 'based on a true story' movie *City of God*, directed by Fernando Meirelles and Kátia Lund, portrays Rio de Janeiro as an urban jungle dominated by poverty, crime and violence in the *favelas*. All of these artists put forth images of cities that did not mesh with the images their authorities would have liked to present. That part of the hypercity which not only conflicts with formal, institutionalized images of the city, but actively seeks to challenge, invert or subvert them, can be termed the shadow city. Batman's 'Gotham City', the dark side of New York, is the archetypal shadow city. But as every city has a hypercity, some more articulated than others, nearly every city has a shadow city, in which urban signifiers combine to form a murkier picture of 'real life'.

Leeke Reinders' contribution *The contra-dance: Rapping and the art of symbolic resistance in the Parisian banlieue* gives a portrayal of such a shadow city. Reinders studies rapping as a social and textual practice based on fieldwork in Sarcelles, a Parisian *banlieue*. This area is the epitome of the anti-suburb, branded in the popular imagination as an area of social problems, many of which are linked to immigrant youth. This group of young migrants, known as *les beurs*, adopt forms of expression – dress codes, graffiti, and in particular rapping – that can be analyzed as a 'hidden transcript'. The young men create a contra-symbolic discourse, in which they can claim a dominant position, inverting the marginalized position that the hegemonic symbolic order accords them. The *beurs* use rap as social weaponry, by which to obtain respect and reputation denied them by dominant society. They demand a place in the urban system, but simultaneously position themselves outside the system. Through an analysis of intangible, discursive signifiers, Reinders places the symbolic language of rap within a context of (a discourse of) exclusion and marginality. In doing so, he exposes the contested nature of urban symbolism and highlights the process of subversion and inversion of the hypercity.

Internationally, established power structures are less movable: it is much easier for global cities such as New York, London and Tokyo to make a symbolic imprint than it is for smaller cities or cities in developing countries.

Research in urban symbolism using the hypercity approach starts by looking at representations and simulations of the city. Such an approach takes urban image and identity as a starting point, which can link back to the 'real' material reality of the city. Studying production and consumption of the hypercity, and possible ruptures between image (representation) and identity (concept of self), can illuminate power struggles over the material city at both the local and the global level and can be the basis for applied studies aimed to promote cities. The contributions by Alexandra Bitusikova, Roelf ter Berg and Dirk Noordman in this volume illustrate this relation between the city's identity and image and its urban symbols.

In her contribution *Brussels as a symbol: Controversial images of the capital of Europe*, Alexandra Bitusikova focuses on the perception of Brussels by its inhabitants as well as the image of the city in a European context. The city struggles between its dual position as the national capital and the capital of Europe. The European perception is that Brussels is not only a boring, ugly city, but a symbol of bureaucracy and corruption as well. But, though the city lacks natural symbolism, the local context reveals symbols important for the urban identity (for example in architecture and in the famous 'Manneken Pis'). European symbolism is found primarily in the European Quarter, which again has the image of a cold, impersonal, bureaucratic, separate area in the city. Moreover, the Eurocrats, who often live in Brussels only during working days and no longer than a few years, do not really identify with the city. European leaders call for more European symbols that could strengthen the position of Brussels as the capital of Europe. However, Bitusikova observes that the lack of European symbols is related to lack of European identity; this European identity cannot be persuaded merely through the creation of symbols. More effective ways must be found to include European citizens and make Europe part of their identity.

In *Selling the city: Representations of a transitional urban community*, Roelf ter Berg discusses the city marketing strategies of Cape Town, from the transitions in 1994 until the present. In the aftermath of these transitions Cape Town joined a trend that also can be seen in other 'wannabe world cities' aspiring to be global players. A Joint Marketing Initiative between public and private organizations was established to create an image of Cape Town to promote the city externally. A positive image of the city was necessary to attract investors and tourists and organize major events and film shootings. This image portrayed the city as a gateway to Africa, a place of historical events and natural beauty, but also presented Cape Town as a cosmopolitan city with high-quality infrastructure, entertainment and shopping facilities. Only recently has the municipality realized that besides external promotion, the city also had to be promoted internally: the Capetonians themselves should be included. Ter Berg shows how this change in attitude is reflected in the changing logos of Cape Town, the most recent one showing people instead of just the natural symbols. Furthermore, the policymakers now also direct attention to the real urban situation, including crime, unemployment and HIV/AIDS, because they realize that their image must be supported by the inhabitants and by reality in order to play a meaningful role in the city's development.

The hypercity approach has more than just theoretical relevance; it can be and is applied intensively in the practice of urban image building. In the context of globalization and the integration of local, regional and sometimes even worldwide city systems, urban identity and image become increasingly important, on the one hand because people and cities want to deploy and enjoy their own distinctive character in a disorderly and uncontrollable global environment, and on the other hand because of the growing competition between cities for employment opportunities and economic development. This type of research is often commissioned by local authorities who want to boost their town in a competitive environment or prevent the loss of important institutions that ensure their city's dynamism.

In *The use of symbols in city marketing: The case of Delft*, Dirk Noordman presents an example of the procedure followed in such applied research. For Delft, a description was made of several dimensions of the city and its community: location, history, size, exterior, interior, symbolism, behaviour and communication. This resulted in a picture of the city's identity and image and provided a possibility to focus on policies enforcing a positive representation. It is important to include both identity and image in the analysis, as a certain level of congruence is necessary to realize the adaptation of perceptions. Consequently, identity and image are always connected, albeit with a certain flexibility. In the case of Delft the image of Delft Blue porcelain, the Technical University and the loss of a number of important institutions over the course of time appear to be important items for reflection. As the city hosts the burials of the Dutch royal family (the House of Orange) in an elaborate, mediatized way, Delft is strongly associated with death and its image is coloured 'black' to a certain extent. In the end, Johannes Vermeer seems to be the most appropriate symbol for the marketing of Delft.

Cityscapes

Research based on the urban symbolic ecology and hypercity approaches described above is mostly of a 'grounded' theoretical character focusing on the development of concepts. It is exploratory and is not based explicitly on specific theoretical questions. One way to advance investigation is to depart from a clear hypothesis. An example is an essay written by Blockmans (2003) on the expression of state power in the architecture of three capital cities, Berlin, Moscow and Istanbul. National leaders use urban symbols to demonstrate their power to their subjects and Blockmans (2003: 7) sees capital cities as 'theatres of state power'. He gives an historical explanation of the link between symbols of state power and capital cities by referring to the rise of nation-states and their efforts to make ordinary people identify with a concept of state that is much more abstract than that of the city or the region, which they were used to. The city can be observed and recognized; people can walk through it and construct a mental image of it. Consequently, it is easier to develop a sense of belonging to a city than to generate identification with a state. This is why the state tries to make itself visible and tangible through the use of architecture and other material symbols in cities, especially capital cities. Moreover, capital cities are a stage for political ceremonies and rituals, festivals and other events in which the state presents itself and shows its power. This implies a role for the capital city that differs from that of other cities, in that the capital functions as a centre of explicit state expression. These cities are employed as strategic symbolic instruments in the nationalist projects propounded by states.

New regimes also use capital cities to present themselves in a symbolic way. Blockmans shows how new regimes generally make use of existing symbols and architecture, such as palaces and cathedrals, to express their power by adapting them to the new ideology. In this way they create a sense of continuity and legitimacy. But new regimes can entail such deep ideological rupture or religious change, the latter demonstrated in the case of fifteenth-century Constantinople, that old symbols are destroyed and replaced by new ones. The way in which states use capital cities to express their power and ideology is not confined to Western cities; the example of Jakarta under Sukarno has been detailed by Nas (1990, 1993). A phenomenon apparent in a number of cities over the past decades is the withdrawal of the state from guiding urban morphology. This development is related to privatization and the integration of cities in the global economy, a process connected to the ideology of capitalism and liberalism; this is where the importance of ideology becomes apparent. Ideology and the need to legitimate the state are important factors in the way cities are used as theatres of state power. To what extent the state can and will withdraw from influencing urban identity and leave the shaping of cities to multinational companies and private investments, as well as the local responses to this phenomenon, remains to be seen (Blockmans 2003).

In addition to capital cities, 'wounded cities' can be marked as a separate type of city. Susser and Schneider (2003) employ this concept in the context of urban symbols as an expression of order and cohesion in a city. They use an organic metaphor to describe the cohesion; when a city fails to serve some of its inhabitants it is seen as malfunctioning. To those persons the city may look like a 'wounded city' (Harvey 2003: 28). Naturally, the use of this organic metaphor is contested. Attention must be given to the city as part of a regional, national and international network as well as to the individuals who constitute it. Yet the organic metaphor can be used analytically to describe 'wounded cities' (Susser and Schneider 2003). The organic metaphor focuses our attention on the various way in which cities may be wounded or malfunctioning. Wounding can happen suddenly, as in the cases of the 9/11 attacks in New York or the bombing of Hiroshima and Nagasaki, but it can also be a process developing during a war or as the consequence of rising crime rates or an economic crisis. An example is Belfast, a city wounded by political violence. Bryan (2003) mentions empty housing, flags, barriers and gates that mark the separation of Catholic and Protestant districts, at the same time being symbols of this separation. Murals and parades are other symbols strengthening the division between the groups.

According to Susser and Schneider (2003) globalization processes can have a major influence on wounding and recovery. Recovery can be a long, hard process, as demonstrated by the case of Beirut with its many displaced persons. Attempts by the authorities to create places and buildings inspired by Western cities such as Paris, London and New York as a way to forget the trauma of the civil war did not resolve the housing problems of the citizens (Sawalha 2003).

An interesting starting point for analysis may be found in important crucial events which cause the city to become wounded. Major examples are the 9/11 attacks on the World Trade Center in New York and the fall of the Berlin Wall. These events did not just change the development of the world community, but were also located in a concrete city, New York and Berlin respectively, and influenced general symbolism as well as local urban symbolism in a major way. The development of Berlin and the filling in of the empty zone of land dividing East and West Berlin by government buildings and by private companies is a promising avenue for urban symbolic analysis. In New York, the prize contest for the design of the new WTC at Ground Zero, the values expressed in the proposals and the drive to construct once more the world's highest building there are as many indications for fruitful urban symbolic study and argumentation.

In addition to these categorizations of cities as 'capital' and 'wounded', urban studies has seen the construction of typologies developed in several ways (Nas 1990a: 62–71): deductively by means of intuitive approaches such as Hannerz' distinction between coke, commerce and court towns (Hannerz 1980), or using historic evaluation, distinguishing generations and families of cities such as Latin American colonial cities or Andean mining towns. Inductive typologies are based on one or a few criteria such as population and density or on several criteria analyzed statistically through principal component or factor analysis. In the context of the previously mentioned concept of urban landscapes, idealized cityscapes can be discerned, including the '*kraton* city', the 'port city', the 'frontier town' and the 'new town'.

Conclusion

This introduction distinguished two broad, complementary approaches to urban symbolism: urban symbolic ecology and the hypercity. Urban symbolic ecology developed strongly over the past fifteen years, mainly on the basis of case studies of (often Indonesian) cities aiming at the formulation of new concepts. In this article the pursuit of new directions and inspiration is expressed and articulated through the new concept of the hypercity. Naturally, general anthropological studies of symbolism remain important in both approaches. These studies consider the ritual and symbolic level of society as a dialogue on social structure and daily life: rituals and symbols both reflect daily life and are comments on it. However, everyday life itself does not function as merely the subject matter; it deals with meanings and incorporates, changes and contradicts them (Nas and Persoon 2003). In addition to semiotics, this general anthropological approach to symbolism remains a root of fundamental research in this field and should be considered a valuable source of inspiration for urban symbolic ecology. The same holds for specific theoretical approaches such as structural functionalism. Other directions indicated in this contribution lie in grounded theory based on further urban case studies. New approaches suggested in this contribution employ existing but underutilized concepts such as 'idealized symbolic landscape', specific theoretical key questions (such as nation building) or crucial social events (for example the 9/11 tragedy). Additionally, the analysis of the development of symbolic systems and the unveiling of one or more dominant themes, or the character or 'soul' of the city are promising to urban symbolic ecology and the hypercity. The possibilities to use the results of these approaches for applied purposes should not be underestimated. In the context of a globalizing world, cities are increasingly integrated into one encompassing hierarchical system, albeit with regional biases, that leads to specialization and occasional fierce competition on the basis of identity and image; concepts that form the cornerstones of urban symbolism. In the context of the widening scale of urban symbolic systems in the world, it is also feasible and necessary to consider the possibility of large-scale quantitative research, which due to the present emphasis on case studies has failed to materialize as yet.

References

Barthes, R. (2000) *Mythologies*. London: Vintage. [Translated from the French by Annette Lavers, original edition 1957.]
—— (2000a) *Elements of Semiology*. New York: Hill and Wang. [Translated from the French by Annette Lavers and Colin Smith, original edition 1964.]
Baudrillard, J. (1993) *Symbolic Exchange and Death*. London: Sage. [Translated from the French by Iain Hamilton Grant, original edition 1976.]
—— (1994) *Simulacra and Simulation*. Ann Arbor: University of Michigan Press. [Translated from the French by Sheila Faria Glaser, original edition 1981.]
Blockmans, W.P. (2003) 'Reshaping cities: The staging of political transformation.' *Journal of Urban History*, vol. 30, no. 1, pp. 7–20.
Bryan, D. (2003) 'Belfast: Urban space, "policing" and sectarian polarization.' In: J. Schneider and I. Susser (eds), *Wounded Cities: Destruction and Reconstruction in a Globalized World*, pp. 251–69. Oxford and New York: Berg.
Colombijn, F. (1994) *Patches of Padang: The History of an Indonesian Town in the Twentieth Century and the Use of Space*. Leiden: Research School CNWS.
Cybriwsky, R. (1997) 'From castle town to Manhattan town with suburbs: A geographical

account of Tokyo's changing landmarks and symbolic landscapes.' In: P.P. Karan and K. Stapleton (eds.), *The Japanese City*, pp. 56–78. Lexington: The University of Kentucky.

Eco, U. (1998) *Faith in Fakes: Travels in Hyperreality*. London: Vintage. [Translated from the Italian by William Weaver, original edition 1967.]

Hannerz, U. (1980) *Exploring the City: Inquiries toward an Urban Anthropology*. New York: Columbia University Press.

Harvey, D. (2003) 'The city as a body politic.' In: J. Schneider and I. Susser (eds), *Wounded Cities: Destruction and Reconstruction in a Globalized World*, pp. 25–44. Oxford and New York: Berg.

Lynch, K. (1975) *The Image of the City*. Cambridge, Mass.: MIT Press.

Meinig, D.W. (1979) 'Symbolic landscapes: Some idealizations of American communities.' In: D.W. Meinig (ed.), *The Interpretation of Ordinary Landscapes: Geographical Essays*, pp. 164–92. New York and Oxford: Oxford University Press.

Nas, P.J.M. (1990) 'Jakarta, stad vol symbolen.' *Antropologische Verkenningen* vol. 9, no. 3, pp. 65–82.

—— (1990a) *De Stad in de Derde Wereld*. Bussum: Coutinho.

—— (ed.) (1993) *Urban Symbolism*. Leiden: E.J. Brill.

—— (ed.) (1998) Special issue: Urban rituals and symbols. *International Journal of Urban and Regional Research,* vol. 22, no. 4, pp. 545–622.

Nas, P.J.M. and M.A. van Bakel (2003) 'Small town symbolism: The meaning of the built environment in Bukittinggi and Payakumbuh.' In: R. Schefold, G. Domenig and P.J.M. Nas (eds), *Indonesian Houses: Tradition and Transformation in Vernacular Architecture*, pp. 461–81. Leiden: KITLV Press.

Nas, P.J.M. and G.A. Persoon (2003) 'Introduction: Signs and symbols.' In: P. Nas, G. Persoon and R. Jaffe (eds), *Framing Indonesian Realities: Essays in Symbolic Anthropology in Honour of Reimar Schefold*, pp. 1–14. Leiden: KITLV Press.

Nas, P.J.M. and R. Sluis (2002) 'In search of meaning: Urban orientation principles in Indonesia.' In: P.J.M. Nas (ed.), *The Indonesian Town Revisited*, pp. 130–46. Münster: Lit Verlag and Singapore: ISEAS.

Nora, P. (1984–1992) *Les Lieux de Mémoire*. Paris: Bibliothèque Illustrée des Histoires. [3 Volumes.]

Sawalha, A. (2003) ' "Healing the wounds of the war": Placing the war-displaced in postwar Beirut.' In: J. Schneider and I. Susser (eds), *Wounded Cities: Destruction and Reconstruction in a Globalized World*, pp. 271–90. Oxford and New York: Berg.

Scott, J.C. (1998) *Seeing like a State: How Certain Schemes to Improve the Human Condition have Failed*. New Haven and London: Yale University Press.

Susser, I. and J. Schneider (2003) 'Wounded cities: Destruction and reconstruction in a globalized world.' In: J. Schneider and I. Susser (eds), *Wounded Cities: Destruction and Reconstruction in a Globalized World*, pp. 1–23. Oxford and New York: Berg.

Chapter 2

'MEGAPROYECTO EL BARRANCO' IN CUENCA, ECUADOR
Architects and their role in local representations

Christien Klaufus

> *In the urban landscape, it is 'El Barranco' of the*
> *river Tomebamba that is preserved with major*
> *consistency in the collective memory as an image*
> *highly representative of the City.*

(Municipalidad de Cuenca, n.d. A, author's translation)

Introduction

Cuenca is Ecuador's third city and one of the country's main tourist destinations. Situated at 2,500 metres above sea level in the highlands of the southern province of Azuay, it has a mild climate and beautiful surroundings with green hills and valley pastures. The administrative canton Cuenca, made up of the city and its rural *parroquias*, has over 400,000 inhabitants of whom about 280,000 live in the city itself (INEC 2003). Thanks to its well-preserved historical inner city, Cuenca was placed on UNESCO's World Heritage List in 1999. Although the city is usually presented as a town with a colonial atmosphere, the buildings now protected by special laws under the UNESCO mandate predominantly date back to the nineteenth century. Two-storey buildings with tiled roofs and decorated facades represent the Cuencan version of French neoclassicism. Only a handful of buildings actually date back to the colonial era before 1830 (Jamieson 2000: 46). Since the UNESCO recognition of the monumental value of the city centre in 1999, the mayor of Cuenca uses the designation fervently to promote his city and to attract more visitors and (inter)national events. Under the pretext of the UNESCO title 'Patrimonio Cultural de la Humanidad', he initiated renovation and restructuring projects of important inner-city areas.

One of the most prominent visual elements in inner-city Cuenca is 'El Barranco', literally the gully or canyon. It encompasses the river Tomebamba with its big boulders and green riverbed, as well as the monumental buildings overlooking the river from the upper bank. The zone stretches along approximately three and a half kilometres from the Otorongo stairs in the northwest to the 'El Vergel' bridge in the southeast of the city centre. The combination of the winding river, the green riverbed and the variegated brick buildings situated above it, gives the area a highly scenic value. Because of the visible skyline the upper bank creates, the river and the adjacent monumental buildings form a unique image that is a point of reference for Cuencanos and visitors alike, and an effective symbol of Cuenca (see Figures 1 and 2). In June 2003 Fernando Cordero, the mayor of Cuenca, decided to organize a national design competition to generate ideas for what was called 'Megaproyecto El Barranco', a project including the river zone, the monumental buildings north of it, a busy avenue on the south side, and some other landmarks in the area (see Figure 3). It was the first time an open competition in urban design was held in Cuenca.

Before it even started, the announcement aroused a debate among local architects. The decision to organize a national competition challenged existing power relations among architects and urban planners. Some established architects felt their monopoly on work for the municipality was at stake, while the architects normally excluded from municipal project assignments considered the competition a unique opportunity to show their qualities. For the mayor – also an architect – the competition could generate ideas on which to base a project bearing his name. He announced the competition to be held from 4 August till 22 September 2003, but the debate continued afterwards. During my stay in October and November 2003, many of my architect friends and acquaintances were involved in the discussion, which made me aware of the different views regarding the design competition and the deeper values of the location involved.

I will analyze in this contribution the discourses and actions accompanying the contested design competition 'Megaproyecto El Barranco' in order to understand the socio-spatial values of urban design in

Cuenca in general, and the symbolic values of this specific area.[1] I address the question of how architects are generally involved in urban design, as well as interpret some opinions on the future development of 'El Barranco', from both participants and non-participants in the design competition. Although the competition was meant to stimulate the discussion of the plans themselves, it was the procedure that was debated most. By analyzing the controversy around the design competition I aim to unravel local architects' stances on urban design, and on the representation of local cultures and social categories in urban space. The conflict shows how in the professional world in Cuenca social status and cultural symbols have become intertwined over the years and how resilient those symbols are to changes.

Figure 1 Tomebamba as part of the 'El Barranco' area (Photograph: Xavier Ordóñez, 2002).

---

[1] This article is based on eight months of fieldwork in Cuenca conducted during three stays over the period 2001–2003. As part of a PhD study in anthropology on the perceptions and practices towards architecture and domestic space in urban lower-class neighbourhoods, I conducted a parallel fieldwork in Riobamba (six months). I interviewed architects, urban planners, decision makers and neighbourhood inhabitants. Because of my own architect background (I graduated with an MSc degree in Architecture in 1993), I soon became involved in the professional circle in Cuenca.

Figure 2 'El Barranco', buildings on the upper bank (Photograph: Christien Klaufus, 2003).

'El Barranco' has an emblematic quality that many Cuencanos associate with the city's identity. Geographically, the river Tomebamba constitutes a significant border line between the upper-bank historical inner city and the lower side of town, 'El Ejido'. The monumental buildings on the upper bank accommodate private housing, cultural institutions, hotels, bars and restaurants. It is the main tourist area in town, and among the inhabitants and many users of 'El Barranco' are foreigners. On the lower side extends the city's first expansion area called 'El Ejido', planned as a 'garden city' in the mid-twentieth century. Today, it contains an upper-class neighbourhood with detached homes, a soccer stadium and modern-style high-rise office buildings.

Parallel to the river runs a big and busy avenue, 'Avenida 12 de Abril', where the campus of the University of Cuenca is located. The riverbanks are connected by several bridges and stairs (to bridge the height between the southern and northern river bank), one of them in front of the main entrance of the university campus. Many pedestrians use the bridges and stairs on a daily basis, but some of the stairways are notorious places for assaults and robberies, and university students often complain about insecurity. A student was robbed and killed a couple of years ago when he went down the stairs on his way to the university. Although 'El Barranco' has its quiet, picturesque and beautiful spots – tourists are often attracted by the picturesque scene of dark-faced women washing their brightly coloured clothes in the river and putting them to dry on the river banks – some parts are socially unsafe and avoided by citizens.

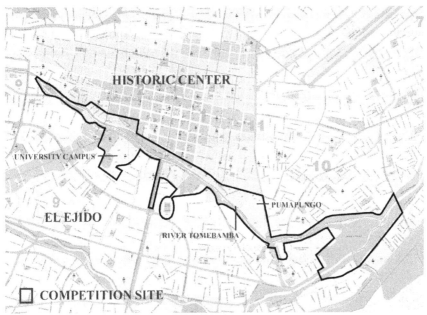

Figure 3 Competition site 'Megaproyecto El Barranco' (Source: Municipalidad de Cuenca, map adapted by author).

Other important landmarks in the area, beside the river Tomebamba itself, are a chapel called 'Cruz del Vado', 'El Paraíso' park, and the nearby archaeological site Pumapungo that contains architectural remains of pre-colonial occupancy by Cañari and Inca people. The Pumapungo site is terraced and highly visible from the lower side of the river. It was recently restored and opened to the public in 2003.

The aim of 'Megaproyecto El Barranco' was to start a development process that might turn this city axis into a major tourist attraction. The proposals for the remodelling of 'El Barranco' had to integrate all the prized places along the riverside, which made this project unprecedented in its extent and impact. To interpret the issues at stake, I will first describe the symbolic meanings of the landscape to the Cuencan upper class and its value in activities to market the city. I will discuss the competition, the disagreement it generated and the role of different groups of architects in the controversy.

Class-based constructions of landscape and past

Elitism and social distinctions in Cuenca, like in many other parts of the Andes, are based on a dichotomy between the city and the rural surroundings. In the dichotomy, the city is associated with mestizo/white upper- or middle-class residents, while the rural surroundings are associated with the mestizo/indigenous popular classes (see De la Cadena 2000; Jones and Varley 1999; Whitten 1981). In Cuenca many elite families present themselves as descendants of the Spaniards (locally identified by their surnames) and claim that they have the culture and knowledge to know what is best for the popular classes in the *barrios* and villages. They are locally referred to as '*los nobles*'. In their class-based geographical construction of superiority, the rural landscape has a double connotation. It is simultaneously associated with the uncultured people that inhabit the rural territory, and with the bucolic values of the scenery: '[i]n opposition to [a] fearsome image, the countryside is [...] seen as terrestrial paradise, the home of peace and tranquillity, and as fostering the most noble human qualities' (Hirschkind 1980: 165).

At the end of the nineteenth and the beginning of the twentieth centuries, local poets and artists from elite families – such as Remigio Crespo Toral and Honorato Vázquez – glorified pastoral life as if it where Arcadia. Literacy and arts being a favourite elite activity, the poets expressed only the perceptions of the upper-class Cuencanos. They glorified their city, blessed with a natural environment resembling paradise, for a reason: 'Cuencanos, and especially Noble Cuencanos, seek to preserve and enhance a distinguished and attractive image for their city, an image which is simultaneously a reflection of themselves' (Hirschkind 1980: 315).

This image of the landscape still prevails, and an eclectic but equally class-based interpretation of the past is included in this projection. Pre-colonial ancestors such as the Cañaris, who lived in this area for more than 500 years before the Incas occupied the settlement (Jamieson 2000: 21), as well as the Incas who came to this region at the end of the fifteenth century, are presented as courageous people who lived in close harmony with nature. The colonial hacienda owners, to which the Nobles trace back their roots, are seen as equally successful inhabitants of the landscape. The urban elite express a paternalistic concern for the scenic maintenance of the rural surroundings. To them, the harmony between the landscape and man-made buildings – sublimated in vernacular house-building practices – has to be protected from 'unauthentic' and 'unethical' influences. This narrative becomes clear in current discussions over so-called rural 'migrant architecture' – opulent houses built by transnational migrants are said to spoil the beauty of the landscape and to ruin the harmony between the city and its natural surroundings (Borrero 2002; Jaramillo 2002; also see Klaufus forthcoming). In the hegemonic discourse, conspicuous villas and consumer goods are of no use to peasants 'who do not know how to use them'. An urban myth that exemplifies this general contempt for peasants is a story of a peasant family living in a newly built four-storey house with an elevator. They use the elevator to store pigs, because they do not know what an elevator is for, or, in its other variant, because there is no electricity in the village. Such narratives underline the social construction of superiority and the paternalistic stance of the urban upper class vis-à-vis the rural inhabitants in the region.

The urban upper class idealizes the bucolic values associated with the countryside because it coincides with their self-image as a contemporary urban aristocracy enjoying outdoor life. Many of the elite are of the class of former landowners and almost all their families have country houses outside Cuenca where they spend their weekends and holidays. The favourite pastime activity is going to a *quinta* (country estate) with relatives and friends to prepare a big meal with roasted meat, drink liquor, dance and relax. As guests are always treated with much hospitality, I have been invited to the *quintas* of many of my Cuencan architect friends. A friend of mine explained to me that upper-class citizens like to idealize rural life and that they appropriate rural elements in their lifestyles to maintain their pretence as today's *hacendados*. Hirschkind already noticed this at the end of the 1970s: 'Agrarian activity is the epitome of appropriate Noble occupations. What it lacks in economic significance as sustainer of the Noble category is compensated by its symbolic significance. No nobler occupation exists for Cuencanos, except perhaps that of poet or priest' (Hirschkind 1980: 214–15).

'Good' elements of rural life and the beauty of nature, combined with such urban qualities as decency and culture, contribute to the elitist urban idyll. The best of both geographical domains are personified in a local symbol, *la Chola Cuencana* (see Weismantel 2003). The Cuencan *Chola* is an archetypical woman of mixed Spanish and indigenous blood, dressed in a colourful *pollera* (multi-layered skirt), an embroidered blouse, her shoulders covered with a shawl, her head covered with a Panama hat and her long black hair worn in two braids. Her qualities are praised in a statue at the entrance of the city and in an unofficial anthem. On almost every official occasion a doll or a person dressed up as *la Chola Cuencana* shows up to represent *'la Cuencanidad'*.

> The *chola* still preserves her original beauty: Creole with fine features, a little bit bronzed, with a clean face, owner and bearer of the autochthonous of her land. She is the developer of the landscape, of the flowers, of the rivers, of the beauty of Cuenca.

> (Editores y Publicistas Cuenca 1990: 264–65, author's translation)

An idealized vision of rustic life and the beauty of nature have thus come to form the basic ingredients of local urban identifications.

Although the *chola*'s rustic qualities are highly appreciated by the elite as well as by the urban middle class, the geographical duality that lies beneath the constructed differences of class, ethnicity and gender is carefully maintained. The real-life urban *cholas* – mostly market vendors – are often addressed with contempt. An example of the unequal position of urban and rural inhabitants of the canton, and the ambivalent position of *la Chola Cuencana*, informs the election of beauty queens. Every year in November during the *Fiestas de Cuenca*,[2] an urban beauty queen and a rural *Chola Cuencana* are elected in the canton. While the urban Miss Cuenca is presented at the front pages of local and even national newspapers, the elected *Chola Cuencana* is presented on page seven only in a local newspaper.[3] In her function of beauty queen, she is not half as influential as her urban sister. Her task is only to 'revitalize and rescue the cultural identity of the women of the countryside' (*El Mercurio* 5 November 2003).

I became familiar myself with the importance of rural symbols in the urban upper-class construction of social difference when I visited a show of thoroughbred Arabian horses in a fancy country estate near Cuenca.[4] The crowd was made up of rich men in casual clothing, accompanied by well-dressed women with expensive sunglasses, jewels and big-brimmed hats that to me looked like cowboy hats. To emphasize the social difference between

---

[2] The annual celebration of Independence Day on 3 November 1820.

[3] See the differences in media attention, for example in *El Mercurio* 5 November 2003, page 7A and *El Comercio* 1 November 2003, pages A1 and D7.

[4] Thoroughbred Arabian horses are generally considered to be the finest and fieriest horses in the world. The owner of the winning horses in that show was a member of one of Cuenca's richest families, Eljuri, who are themselves of Arabian origin but now belong to the local upper class.

the urban visitors and 'real' rural inhabitants, the organization had hired a folklore group of six female *chola* dancers, dressed in the typical outfit, who danced barefooted on the grass during the break. Their folkloric dances were met with loud applause, and yells and whistles from the audience. Yet to me the yells and whistles coming from the tribune sounded exaggerated, and I did not know if they were really appreciating the show or just making fun of the six barefooted girls hired to dance for them. I see the symbolic value of *la Chola Cuencana* as situational and opportunistic: at one occasion she is presented as the main symbol of *Cuencanidad* thanks to her representation of what is considered autochthonous and natural, while at other times the symbol of the *Chola Cuencana* is used to represent only the lower-class rural segment of the region.

Identification with pastoral life not only encompasses the surroundings of Cuenca, it also absorbs landscape elements in the city itself, which are said to form an inextricable part of local urban identification:

> With the landscape we form a kind of unbreakable union, manifested in our actions and reactions, in our social order and in the artistic modality. A reflection of this landscape, serenely insisting in the equilibrium and the modesty, are the sobriety, the courtesy and gravity of the Cuencano; his individualism that is not tough on solidarity; his love for exactness and decorum. ... *The Cuencano does not live in the landscape: he lives the landscape*, like a snail the tiny home that he lifts upon his shoulders.
>
> (Jara Idrovo 1998: 20, author's translation and italics)

The narrative about the traditional harmony between culture and nature, valued as an idiosyncratic Cuencan characteristic, applies to 'El Barranco' in a similar way as it does to the valleys, hills, villages and country estates surrounding the town, with only one difference: 'El Barranco' has a unique emblematic quality that also serves to promote the city.[5]

The symbolic economy of Cuenca
Preceding his second term of office (2000–2004), Fernando Cordero, architect and mayor, formulated fourteen objectives as part of his local development plan. One of the objectives mentioned was to reinforce the city's image in order to attract more tourism, thereby strengthening the local economy (Municipalidad de Cuenca 2003). Although the city already

---

[5] See the municipality's description at the beginning of this article.

attracts many (inter)national tourists who customarily visit the historical centre, the National Park 'El Cajas' on the western side of town, and are drawn by the local Panama hat industry, Cordero, intended to entice more visitors. In order to promote his city and to compete with other cities – and with other mayors – he had to exploit the city's potential tourist sites such as 'El Barranco' to market Cuenca.

A famous recent example of successful city marketing in Ecuador is the prestigious project 'Malecón 2000' in Guayaquil. In a public–private partnership, a foundation has managed to change a rather run-down seafront area into a commercial zone with a cosmopolitan atmosphere resembling Miami. With an estimated area of twenty hectares, the newly developed project includes museums, green areas and commercial centres. The objective, in which they succeeded, was to revive the area economically by attracting more tourists (Municipalidad de Guayaquil n.d.). The project was started in 1998 and finished in 2002. In the media, 'Malecón 2000' is considered a successful example of urban regeneration and gentrification (in the sense used by Jones and Varley 1999) and in 2003 the United Nations rewarded the mayor of Guayaquil with a prize for good local governance (*El Comercio* 4 December 2003). Generally speaking, 'Malecón 2000' seems to have become an urban emblem with national allure that made many Guayaquileños proud of their city again.

More than anywhere else, the two successive mayors of Guayaquil involved in the project[6] have managed to strengthen their own governing power as well as local identity – feeling Guayaquileño – by using what Zukin calls the symbolic economy of the city:

> [T]he symbolic economy features two parallel production systems that are crucial to a city's material life: the *production of space*, with its synergy of capital investment and cultural meanings, and the *production of symbols*, which constructs both a currency of commercial exchange and a language of social identity. Every effort to rearrange space in the city is also an attempt at visual re-presentation.

(Zukin 1998: 23–4)

In the symbolic economy of the city, urban space is embedded with codes of inclusion and exclusion, and it has broader potential to sell a (desirable)

---

[6] During the time that 'Malecón 2000' was designed and constructed, Léon Febres Cordero and Jaime Nebot were the successive mayors.

urban image. Inner-city reconstruction projects in other cities of the world have proven to generate a variety of commercial activities and cultural meanings (Ellin 1996: 63–7); therefore they are prioritized in many Ecuadorian government plans, and not only in Guayaquil.

In Cuenca, Fernando Cordero similarly understood the commercial and social values of Cuenca's own local symbols: the representation of a glorious past and of a bucolic landscape. He used these a few years ago when he decided to remodel the city's central park, 'Parque Abdón Calderón'. Since inner-city Cuenca and its neoclassical architecture had been acknowledged by UNESCO as Cultural Patrimony of the Humanities, the presumed harmony between nature and culture became a symbolic currency that could now be exploited legitimately, although it was not done openly. The reconstruction work in the park was executed behind a wall of wooden plates that shut off the park form the public view for about a year. When the work was done and the park reopened in 2003, it reappeared as a copy of its nineteenth-century neoclassical original (see Figure 4). It was laid out in geometrical forms that shaped green areas and walkways. The park was decorated with a fountain, a statue in remembrance of Abdón Calderón and a neoclassical kiosk that resembles the original one. In nineteenth-century Cuenca, French neoclassical architecture was the style with which the self-ascribed aristocrats identified most (Espinosa Abad and Calle Medina 2002). As such, the architectural image of the original plaza was strongly connected to the social hierarchy in the nineteenth century. By copying the design, the inherent hegemonic representation of social life was also included in the new layout (see Low 2000). Just like its nineteenth-century version, the new park came to represent the cultural values of Cuenca's elite.

Under the pretext of the city centre as a World Heritage site, the municipality initiated several gentrification processes. With the slogan 'a window on the past, a door to the future' the mayor stressed the importance of (constructed) traditions for city-marketing purposes, while he equally recognized the need for modern public services in order to compete with other cities in a globalizing world (Municipalidad de Cuenca n.d. B, author's translation).

The river zone being an important landscape element in town with the capacity of becoming a main recreational area, the development of 'El Barranco' seemed another logical target of Cordero's politics. Not only because he wanted to impress the public as a mayor and an architect, but also because this area exemplifies the hegemonic way of framing Cuenca as an extraordinary environment where nature and culture coincide. He aimed to

Figure 4 Parque Abdón Calderón (Photograph: Christien Klaufus, 2003).

attract more tourist capital by reinforcing the natural image of the area and by using the riverside as a place for modern leisure activities, instead of as a washing place, the location of assaults and robberies, and the polluted transportation axis that it is right now. The reconstruction could also be used indirectly as social politics to get rid of 'unacceptable' spatial functions and 'unwanted' users, and instead create a social and spatial order that excludes certain groups of people (see Jones and Varley 1999). Whether or not this is going to happen remains to be seen, as the project is still being developed. 'El Barranco' being the emblem of Cuenca, a discussion on how local values can be reinforced and expressed architecturally for the public good was to be expected anyway, but the design competition gave it a major impulse. As I mentioned in the introduction, it was especially the procedure of the competition that aroused the most debate; therefore, I will now sketch the role of local architects in contemporary urban design projects.

Local architects

As described above, Cuenca is a city with a small but rather dominant group of elite families, many of whom claim to be the descendants of Spanish colonial nobility. This elite minority controls the allocation of resources and the flow of information, and they occupy many important positions in the municipal, commercial and cultural institutions (Hirschkind 1980; Lowder 1989). Architecture and urbanism are primary domains over which they exert control. The dominance of the spatial and physical domain dates back to the mid-twentieth century when the old land-owning elite started to sell off parts of their land as a result of agrarian reform laws and a changing local economy.

During that time, the land-owning upper class considered it opportune to have urban planners and architects among their relatives, so that they could control all stages of urbanization. They pressed the University of Cuenca to open a Faculty of Architecture, which started to function in 1961 (Lowder 1990). At the end of the 1960s, Cuenca had its first generation of locally trained architects and urban planners. Some of them became famous for their local architectural style, called 'la Arquitectura Cuencana', stone villas with tiled roofs. They tried to combine in this style modern techniques with locally produced materials such as bricks, natural stones and tiles, in order to create a regional style in accordance with the landscape. Their clients were upper-class citizens. The first generations of locally trained architects, who graduated in the 1970s and 1980s (mayor Fernando Cordero was one of them) have a good professional reputation both locally and nationally.

Many Cuencan architects combine the work in their own offices with jobs in local institutions such as the Faculty of Architecture, the 'Casa de la Cultura', the commercial chambers and the municipality. As Hirschkind describes, multi-job holding is a common strategy among elite members to safeguard their dominion over local resources:

> Multiple job-holding, or multiple established interests, is another mechanism used by Nobles in resource competition. In an established pattern, Nobles occupy several paid positions and a variety of unpaid but influential or prestigious ones, all at one time. More than any economic benefits such as a schedule might bestow, this pattern provides Nobles with extensive and intimate contacts in a number of social and economic spheres, contacts which may be useful as allies, as collaborators on projects of mutual interest, or as resources of information. ... This pattern of direct control from the top of local formal institutions helps to protect and guide the channeling of resources according to Noble convenience. It assures at least a consideration of Noble interests in almost all areas of formally constituted authority.

(Hirschkind 1980: 137–8)

This pattern of professional job holding amounts to an old-boys network of architects, who divide big design projects among themselves without interference from outsiders. They still sustain their social network of strongly intertwined relationships, in which favouritism is not shunned (Lowder 1989; Miles 1994).

As a result of the interwoven relations in the professional network of architects of this generation, architects from the municipality and the Faculty of Architecture often work together on an informal basis. When the municipality needs a study of a specific urban area, the mayor or the head of the Planning Department pays the Faculty of Architecture to conduct the study or to make the design. Senior architects in the faculty hierarchy divide the irregular jobs and the corresponding earnings in rotation. The studies are carried out by the students participating in their workshops. Out of the student's designs they select the best work, which is then offered to the municipality. Several architects and students told me that the privileged architects working on such a commission keep a substantial part of the earnings, instead of donating it to the faculty.

This practice had gone more or less unnoticed until the beginning of 2003, when the mayor of Cuenca and the dean of the faculty officially signed an agreement promising to give several important municipal urban projects to the faculty. This agreement became public when it was published in a two-year report of the faculty (Universidad de Cuenca n.d.). Among the projects was 'Megaproyecto El Barranco', certainly the most prestigious project promised to the faculty. The subsequent decision of the mayor not to give this assignment to the faculty, but to open up a competition instead, was germane to the controversy I will describe below.

First, it is important to notice that the contemporary professional domain in Cuenca is much more diverse than the relatively small network of elite architects described so far. While this powerful group of established architects still dominate important positions in different institutions, they make up a relatively small part of all professionals working in architecture and urbanism today. In total, a thousand architects are said to be working in Cuenca.[7] Many of them come from less-privileged families. Over the last decades, the Faculty of Architecture changed from an expensive elite study into a career that has opened its doors to students from broader layers of society.

This process of change is partly related to alterations in the university policy itself and partly to social changes in Cuenca's society. Due to massive transnational migration and subsequent incoming remittances, a new class of economically well-off people has been formed. Prosperous migrant families have come to dominate the local economy.[8] As a consequence of this new spending power, over the last decade more families were able to send their children to university, and the formerly exclusivist study of architecture has now come within the reaches of the urban middle class.

Young architects who graduated over the last decade try to survive professionally without necessarily working their way up in the closed old-

---

[7] Interview president Colegio de Arquitectos 29 October 2003.
[8] See Jokisch and Pribilsky (2002) for an overview of the effects of emigration in Ecuador.

boys network, although the competition is strong. They look for niches in the design and planning market and benefit from new groups of clients, especially among migrant families who became important generators of the local construction industry. Many of the younger architects also have their *palancas* (social connections) in different institutions – as such there is no clear-cut division between the two groups or generations – but there seems to be a tendency among the recently graduated architects to open up the closed system of project assignments.[9] One way to accomplish this is through the occupancy of influential positions in the local department of the 'Colegio de Arquitectos' (College of Architects).

The College of Architects is a national professional association that controls the work and functioning of architects. It is organized in regional departments and the College of Architects for the province of Azuay is located in Cuenca. Based on a national law, all practicing architects are legally bound to be registered at one of the regional departments of the College of Architects. The organization has a double task: it guards the quality of the professional practice (e.g. fair competition, minimum honoraria) and it checks if submitted project plans comply with the official norms and regulations. As a result of the active participation of some younger, reformist architects in 2003, for the first time in the history of the College of Architects in Azuay (founded in 1978), the president, Marcelo Astudillo, was not in any way related to the Faculty of Architecture and therefore more independent of the old-boys network than his predecessors. He, too, played a decisive role in the design competition controversy, as I will show below.

The contested design competition

When the agreement between the Faculty of Architecture and the municipality to make an urban design for 'El Barranco' became public, the College of Architects urged the mayor to reconsider his decision. The president of the professional college, Marcelo Astudillo, asked Fernando Cordero to organize an open design competition for architects instead of giving it away to the faculty's architects. He explained to me:

This is a struggle that the College is taking up, because it is not fair that professors making use of, or under the guise of, the Faculty of

---

[9] Based on observations, conversations and professional meetings with about twenty architects of a younger professional generation who graduated over the last decade and now work individually.

Architecture, pretend to have earned the work that should have been awarded through competition including all architects, not just in the region, but all over the country. So, that is the quarrel of the College. We are fighting this agreement, and it seems that we managed to get through. So, we are succeeding in disrupting a little bit this mechanism that they have to obtain work. And it has to be said openly, because they are professors of the university who, under the flag of the faculty, are getting work without competition, especially big and representative projects that at a certain moment can become landmarks for the city. So that is incredible; that is what we are concerned about.[10]

Officially, the mayor was not obliged in any way to organize a competition, because the law states that only institutions without a planning department are forced to contract out projects by means of a competition. Since the municipality has its own planning department, the mayor could easily have ordered his department to make a plan. Besides, as mayor, he has the power to assign projects to whomever he deemed appropriate, and as an elite architect, he had a strong social network of people who supported the patronage system. On the other hand, he was an ambitious politician who wanted to be re-elected in 2004, and the realization of a local development project with national significance might serve that goal. So he decided to organize a competition. The College of Architects triumphed and stated that 'this vigorous Cuencan initiative, tending to democratize the planning and appearance of the most emblematic urban sector of the historic city' was a watershed in local planning procedures (CAE-Azuay 2004: vii).

When Cordero and the College of Architects announced the upcoming design competition, the faculty protested fiercely. On 4 August 2003, the mayor officially opened the National Idea Competition with 22 September as the deadline. One of my architect friends, active in the College of Architects as well as in the Faculty of Architecture, told me the mayor presumably selected this seven-week period on purpose, because the university was closed for summer holidays during that time. This would force the faculty professors to decide whether or not to participate *as professionals* instead of trading student plans for money. The quarrel was fought out in the local newspaper, *El Mercurio*.

The faculty professors organized their critics in two separate initiatives. The directors of the faculty decided to refrain from sending a representative to the jury of the competition, as was agreed earlier. The dean communicated

---

[10] Interview 29 October 2003.

that the faculty, as a member of the Committee of the Historical Centre, stuck to the opinion that previous to the competition the municipality should have made a Management Plan for the Historical Centre. As part of that management plan a Special Urban Plan for 'El Barranco' should have been made also. As this was not the case, the faculty abstained from participating in the jury (*El Mercurio* 24 September 2003).

The mayor then asked the faculty staff to reconsider this decision, asking them if they would also consider it a violation of the law if the mayor asked the staff and their students to develop projects for other urban areas, or if they were asked to form part of a multidisciplinary team that would further develop the ideas stemming from the competition (*El Mercurio* 27 September 2003). By referring to the possibility of channelling future projects directly to the faculty as the municipality used to do, the mayor pressured the faculty not to thwart his ideas.

Another opposing group was formed by six individual architects, some of them professors, others their architect friends joining the group in their protest. They declared that the College of Architects had been wrong to support this initiative, because the municipality did not comply with the Regulations of Architecture and Urban Design Competitions, which states that an Advisory Committee has to be formed. Without having installed such a committee, the competition lacks a connection between the municipality and the jury. Besides, the College of Architects itself was said to have expressed doubts during a round-table conference earlier that year.

The competition was based on the mayor's own plan for 'El Barranco' that he had made twenty years ago. The six opposing architects considered the mayor's insistence to use his own design as a basis for the competition an unethical one. They also thought the competition violated the Law on Intellectual Property because the winners would not be able to execute their plans, which instead would be handed over to a foundation that would continue to work out the design. In the local newspaper they urged the municipality to stop the competition (*El Mercurio* 19 September 2003). The mayor responded in the same newspaper that he prohibited this rebellious group from using the municipality's letterhead or, to put it clearly, he forbade them to represent the city of Cuenca, complaining that they did not serve the local interests at all, since the only thing they defended were their own economic interests.

One day, the president of the College of Architects received an anonymous fax in which the authors criticized his decision to support the competition. The tone of the letter was very personal and the president felt highly offended. The fax appeared to be not as anonymous as the authors had probably hoped for, because the letterhead revealed the name of the location where the letter was sent: the Faculty of Architecture. The president of the professional college complained about the incident both to the dean of the faculty and to the rector of the university. This led to an internal controversy in the university itself and the rector urged the faculty staff not to use the equipment for 'private' matters. The sub-dean was quick to visit the president of the college and to convince him that he was not involved in this protest. Other faculty professors individually tried to dissociate themselves from the protests and tried hard to restore their *palancas* inside the professional college, while the six architects tried to reinforce their opposition by looking for support among architect friends.

Of the opponents of the competition, it was the established architect and renown critic Oswaldo Páez who most actively expressed his opinions. He published an article in which he criticized the competition. According to him, the competition was used as a smokescreen to hide the fact that the mayor already had his own design plan ready to be executed. He further postulated that 'El Barranco' became a local emblem precisely because it is an area that was never really planned by architects, but had simply evolved:[11]

> As El Barranco [consists of] an architecture [made] without architects, constructed from the inside and towards the inside, as it is an architecture of coincidences and contingencies, the place has been converted in a space that makes a metaphor of what could be a human place, a place that suggests that it could be a space of the re-encounter with ourselves, different from the non-places that peer at the non-encounters, the solitude and the dissatisfaction of consumerism.
>
> (Páez Barrera 2004, author's translation)

He considered the mayor's ambition to make the area into a tourist attraction a threat, because it might privatize and commercialize this emblematic area, and convert it into a spectacle. To him, 'Megaproyecto El Barranco' could be compared to 'Malecón 2000': both projects were forms of *'urbanismo de los globalizadores'*, meant only to commercialize the city's landmarks. For him, the symbolic economy of Cuenca was going to be

---

[11] Although Páez is probably right to point to the more or less spontaneous development of the area, architects have been and still are involved in construction activities in the area, so the expression 'architecture without architects' is not literally correct.

exploited in an unacceptable way. His critique was mainly oriented toward the mayor and his city-marketing strategies, but since he did not participate in the competition he did not offer any visual support of his alternative view.

At the close of the competition in September, twenty-six teams involving more than 200 participants handed in their ideas for the future development of the area. Among the participants in the different teams were five university teachers, three of them established architects. The jury, composed of representatives of different local institutions, selected three winning plans. Surprisingly, one of the winning teams, headed by Mauricio Morreno, belonged to an office in Quito. As Mauricio Morreno is a Cuencano living in Quito, nobody expected him to win because nobody believed the jury would let an 'outsider' win. The fact that he had won, seemed to underline the jury's integrity and the national scope of the competition. The municipality organized an exhibition in a local museum where the plans were presented and some months later the College of Architects published a book containing the plans (CAE-Azuay 2004). A foundation, in which several representatives from the city's main institutions participate, comparable to the foundation that managed the 'Malecón 2000' project in Guayaquil, was established to decide how to continue with the development of 'El Barranco'. The foundation could use the submitted plans as sources of inspiration for further development.

The submitted plans

The differences in opinion on the procedure aroused so much debate that a thorough discussion of the submitted plans – for instance which, and whose, local values were represented and how – was relegated to the background. I will try to summarize the twenty-six proposals, taking the four prerequisites of the competition as a point of reference.

The prerequisites were: (1) the valuable architectural patrimony had to be respected; (2) the rich topography had to be utilized to generate socio-cultural activities (which were not specified); (3) the problematic traffic flow on the busy avenue '12 de Abril' had to be restructured; and (4) the plans had to contribute to a reinforcement of the local cultural identity (singular and unspecified).

The first two criteria were generally taken quite literally. Many proposals included the planting of new trees, the construction of new bridges across the river, and the design of new public spaces decorated with fountains, statues and sculptures. The third criterion, the problem of the chaotic traffic flow, was resolved in most plans by remodelling existing roads so that they prioritized pedestrians, and by creating new ones. All three winning teams proposed to convert the busy avenue into a pedestrian zone where motorized traffic was restricted or even prohibited. Winning architect Manuel Contreras stated that this drastic measure would be necessary 'because of the permanent deterioration of the historical and environmental values, due to the intense motorized traffic, incompatible with the geometrically required characteristics' (CAE-Azuay 2004: 2). Also, trams and trolley buses were often mentioned as a solution that combined public transport with a tourist attraction.

The fourth criterion, a nature–culture symbiosis, was used in many plans, most clearly so in the proposal of Salvador Castro who based his plan on 'the nature, which from the edges of the earth, and the culture, which from the edges of history, bring us their heritage' (CAE-Azuay 2004: 22). Not only in his plan but in most proposals, culture was interpreted as a mixed heritage from pre-colonial Cañari–Inca and colonial legacies. Nostalgia for a noble past and an idealized natural environment were thus repeated in the proposals. Remarkably, none of the participants described a local cultural identity as a social construction by and for today's people of Cuenca.

Most teams interpreted 'identity' as the formalistic reproduction of historical elements, such as Inca walls or colonial-style buildings and bridges. They did so without elaborating on the question of whom these monumental spaces represented and for whose daily use they were meant. The team of young architect Xavier Corral, for example, designed a bridge with a glass wall, decorated with words in two languages: Cañari and Spanish. The words in Cañari represented the roots of contemporary Cuenca, whereas the translation of those words in Spanish linked the past to the present society. Not by coincidence, the proposed bridge would lead to Pumapungo Park, where traces of the Cañari and Inca cultures were found. His motivation for the design was:

> a profound respect for the legacy of our ancestors, for the grand masters of the past. Tradition is not something that we can avoid but something that we have to search profoundly [to find] its roots, the revolution in Architecture and in the Environment is the reencounter with things that were lost and that need to be rescued.

> (CAE-Azuay 2004: 26)

Even though some indigenous people in the rural areas of the Cuenca region speak a contemporary variant of the Cañari and Inca languages, i.e.

Quichua[12], and although contemporary Spanish in the Ecuadorian highlands is punctuated with Quichua words, this architect referred to the Cañari and Inca languages as things that were lost and recovered, thereby ignoring the contemporary Quichua-speaking citizens as possible users of urban space.

In Corral's proposal, as in most other plans, the people for whom the new 'El Barranco' was designed seemed to be taken for granted. No *Cholas Cuencanas*, no indigenous groups, no migrant families or other social categories were referred to, even though in many parallel debates on the future of Cuenca the variegated socio-cultural categories often come up. Although two teams of young architects actually did criticize idealized formalistic solutions that lacked correspondence to the contemporary Cuencan society, most participants reiterated the hegemonic narrative of the nature–culture harmony, with 'culture' being the pre-colonial and colonial heritage, not the contemporary indigenous or *chola* culture nor the rather conspicuous forms of consumption of migrant families that characterize today's Cuenca.

Ultimately, the form of the competition appeared to be more innovative than the content of the proposals. The mayor and the College of Architects aimed at a more democratic participation in the plans for 'El Barranco' and they partly succeeded in that by opening up the design process for other professionals than the closed elite circle. The young architects wanted their share in the field of urban design, and the competition offered them that opportunity. However, in their plans for a remodelled 'El Barranco' they used the same hegemonic symbols as did the established elite architects in other urban design projects, for instance in the remodelled 'Parque Calderón'. Repeating Zukin's assertion that in the symbolic economy of the city the production of space and the production of symbols are intertwined, it is clear that in this case, the production process changed but the proposed spaces and symbols did not.

Conclusion

The first open design competition in Cuenca for 'Megaproyecto El Barranco' was accompanied by discussions and conflicts. As the design of urban space and local symbols used to be in the hands of a few privileged architects, belonging to urban upper-class families, they felt their privileged position was at stake. Their dominance in urban design and architecture dates back to the 1960s when land-owning families instigated the foundation

---

[12] Quichua is the name of an ethnic group and its language, which forms part of the Andean Quechua culture.

of the Faculty of Architecture. Their dominance in the professions of architecture and urban design coincided with socio-economic power over the social use of space and over local representations. These architects were used to functioning in an old-boys network in which they occupied several functions in different local institutions at the same time. The intricate social network contributed to a closed system of architectural project assignments. That system was affected by the open competition. The public debate about whether or not a competition should be held exposed the annoyance of the architects who used to work on municipal design projects without having to compete with other colleagues. They boycotted the design competition, officially because the laws were not respected, but unofficially because the mayor had previously promised this project to some of them. The controversy over the competition for 'El Barranco' thus revealed the socio-political and cultural motivations of the professionals normally involved in urban design.

Younger generations of architects were glad to participate in the design competition, because it was an opportunity to show their professional qualities instead of working their way up in a system of favouritism and patronage. One of the young, ambitious architects told me he doubted whether the mayor would really respect the authorship of the winning architects, but he preferred to express his criticism of the urban planning system through active participation – by contributing to the contents of the plans, instead of boycotting just the form of it.

The mayor played a double role in the controversy. He organized the competition instead of asking his colleagues at the faculty to submit a plan, as he had originally promised them. But the open competition was not completely without constraints either – he did not guarantee authorship to the winning teams, so that the foundation could use the plans as think-tank material for future development. In his career as an architect, he had once made a plan that was now integrated in the competition's assignment. As a politician he saw the commercial possibilities of the 'El Barranco' site, which could be exploited by using the symbolic values of the landscape and the local cultural heritage. Being both a member of an architectural elite and an ambitious politician, his ambivalence was reflected in his actions.

Although the procedure of the competition was innovative, the proposals of the participating teams were generally a continuation of a class-based view of urban space and on the symbolic values of urban design. In the elite discourse, a nostalgic appreciation of the pre-colonial and colonial past and a perceived harmony between the built and the natural environment prevail. The symbiosis between nature and (past) culture are presented as a set of generic local values that have to be cherished. The architectural proposals in the competition for 'El Barranco' showed a continuation of this hegemonic vision of Cuenca. The values of the nature–culture symbiosis continued to be expressed by a younger generation of architects. Except for some functional and formalistic alternatives, most plans did not explore the question of contemporary cultural identities, and the new participants in the field of spatial and symbolic production did not produce new kinds of spaces or symbols.

As an epilogue to this case, the impact of the new procedure in the long run can also be questioned, as some events that occurred at the end of 2003 suggest that the opening up of the social system did not last very long. At the end of that year, representatives of the municipality and the faculty came together officially to start a dialogue to solve the 'differences in opinion' between both institutions, to safeguard the collaboration on future projects (*El Mercurio* 18 December 2003). Hence, it can be presumed that the old-boys network has partly been restored and that the established architects have temporarily closed their ranks again. To conclude, occasional developments like the one described here might signal a tendency toward social change, but for now Cuenca's hegemonic forces in the spatial and symbolic economy have prevailed.

References

Borrero, A. (2002) 'La Migración: Estudio sobre las remesas de divisas que ingresan en el Ecuador.' *Universitas, Revista de la Universidad Politecnica Salesiana del Ecuador*, vol. 1, no. 1, pp. 79–87.

CAE-Azuay (Colegio de Arquitectos del Ecuador, Núcleo Azuay) (2004) *Concurso de Ideas sobre El Barranco, Cuenca-Ecuador.* [Includes cd-rom.]

De la Cadena, M. (2000) *Indigenous Mestizos: The Politics of Race and Culture in Cuzco, Peru, 1919–1991*. Durham: Duke University Press.

Editores y Publicistas Cuenca (ed.) (1990) *El libro de Cuenca III.*

El Comercio (2003) 1 November, 'Rebeca Flores es la nueva soberana de los Cuencanos', pp. A1, D7.

——— 4 December, 'El ONU entrega oficialmente su reconocimiento a Guayaquil', p. D1.

Ellin, N. (1996) *Postmodern Urbanism*. Cambridge and Oxford: Blackwell.

El Mercurio (2003) 19 September, 'Barranco origina polémica', p. 8A.

———24 September, '26 trabajos intervienen en el Concurso de ideas', p. 7A.

———27 September, 'El Barranco, solicitan reconsiderar decisión', p. 7A.

———5 November, 'Tania Tapia fue electa como Chola Cuencana 2003–2004', p. 7A.

——— 18 December, 'Municipio y Universidad', electronic document. [www.elmercurio.com.ec, accessed 18 December 2003.]

Espinosa Abad, P. and M. Calle Medina (2002) *La Cité Cuencana: El Afrancesamiento de Cuenca en la Época Republicana (1860–1940)*. Cuenca: University of Cuenca. [MA thesis, Faculty of Architecture and Urbanism.]

Hirschkind, L. (1980) *On Conforming in Cuenca*. Madison: University of Wisconsin. [PhD thesis.]

INEC (Instituto Nacional de Estadística y Censos) (2003) *Censo 2001*, www.inec.gov.ec. [Electronic database.]

Jamieson, R. (2000) *Domestic Architecture and Power: The Historical Archeology of Colonial Ecuador*. New York: Kluwer Academic and Plenum Publishers.

Jara Idrovo, E. (1998) 'El paisaje Cuencano: Dialogo entre el hombre y la naturaleza.' In: *Cuenca de los Andes*, pp. 18–22. Cuenca: Ilustre Municipalidad de Cuenca and Casa de la Cultura Ecuatoriana, Núcleo del Azuay.

Jaramillo, D. (2002) 'Globalización y cultura: Del hogar a la casa fetiche en la arquitectura popular azuaya.' *Universidad Verdad, Revista de la Universidad de Azuay*, vol. 27, pp. 185–95.

Jokisch, B. and J. Pribilsky (2002) 'The panic to leave: Economic crisis and the "New Emigration" from Ecuador.' *International Migration*, vol. 40, no. 4, pp. 75–101.

Jones, G. and A. Varley (1999) 'The reconquest of the historic centre: Urban conservation and gentrification in Puebla, Mexico.' *Environment and Planning A*, vol. 31, no. 9, pp. 1547–66.

Klaufus, C. (forthcoming) 'Globalization in residential architecture in Cuenca, Ecuador: Social and cultural diversification of architects and their clients.' *Environment and Planning D*.

Low, S. (2000) *On the Plaza: The Politics of Public Space and Culture*. Austin: University of Texas Press.

Lowder, S. (1989) 'The distributional consequences of nepotism and patron–clientelism: The case of Cuenca, Ecuador.' In: P. Ward (ed.), *Corruption, Development and Inequality: Soft Touch or Hard Graft?* pp. 123–87. London: Routledge.

—— (1990) 'Cuenca, Ecuador: Planner's dream or speculator's delight?' *Third World Planning Review*, vol. 12, no. 2, pp. 109–30.

Miles, A. (1994) 'Helping out at home: Gender socialization, moral development, and devil stories in Cuenca, Ecuador.' *Ethos*, vol. 22, no. 2, pp. 132–57.

Municipalidad de Cuenca (2003) *Mensaje del Alcalde de Cuenca Arq. Fernando Cordero Cueva a Ilustre Consejo Cantonal a Manera de Informe.* [http://www.municipalidadcuenca.gov.ec/nuevaciudad/inforlab/enero2003/InfLabEne200 3.pdf, accessed 21 April 2004.]

—— (n.d. A) *Cuenca, Candidata a Patrimonio Munidial de la Humanidad: Sintesis de la Propuesta para la Inclusión del Centro Historico de Cuenca, Ecuador, en la Lista de Patrimonio cultural*. Gobierno Local 1996–2000.

—— (n.d. B) *Reinauguración del Parque Calderón: Una Ventana al Pasado una Puerta al Futuro*. [www.cuenca.gov.ec/reporta/ParqueC/Reinau.htm, accessed 6 May 2003.]

Municipalidad de Guayaquil (n.d.) *Malecón 2000*. [www.guayaquil.gov.ec/index. php?idsec=106, accessed 12 May 2004.]

Páez Barrera, O. (2004) 'El Barranco para la multitud. Otra propuesta.' *Trama* 85. [www.trama.com.ec/T85/t85g.htm, accessed 6 May 2004.]

Universidad de Cuenca (n.d.) *Informe de la Facultad de Arquitectura: Febrero 2001–Enero 2003*. Cuenca: Facultad de Arquitectura.

Weismantel, M. (2003) 'Mothers of the patria: La chola Cuencana and la mama negra.' In: N. Whitten Jr. (ed.), *Millennial Ecuador: Critical Essays on Cultural Transformations and Social Dynamics*, pp. 325–54. Iowa City: University of Iowa Press.

44

Whitten Jr, N. (1981) 'Introduction.' In: N. Whitten Jr (ed.), *Cultural Transformations and Ethnicity in Modern Ecuador*, pp. 1–41. Urbana, Chicago and London: University of Illinois Press.

Zukin, S. (1998) *The Cultures of Cities*. Malden and Oxford: Blackwell.

Chapter 3

THE ITALIAN SKYLINE
Family towers as elements in urban symbolism

Gaby van der Zee

Introduction

Its skyline is an important element through which the city has profiled itself throughout history. On royal seals in the Middle Ages, the city was often represented with a prominent place reserved for the principal and often tallest buildings. Even now, fridge magnets, glasses, t-shirts and postcards of city views are sold in large numbers. The skyline and buildings that form it compose the image of the city.

In Italy, the family towers are the most outstanding characteristic of the skyline. As skyscrapers define a city scene in modern times, these towers performed the same function for medieval towns. My first step in this essay to describe the Italian skyline will be to examine what the concept of skyline exactly means and what the characteristics of different skylines are. Then I shall turn specifically to the Italian skyline and the role family towers play and have played in it. The history and the external features of the towers, the institution of tower-associations in which various families or clans were united, and the various threats to the towers will be discussed. More empirically I shall look at two of these family towers in Bologna: the Asinella and the Garisenda Towers. In the conclusion, a comparison will be made between the modern skyline and that of Italian towns in the Middle Ages.

The urban skyline

Spiro Kostof (1991: 279), who dedicates a whole chapter of his masterpiece on cities to the skyline, says the traditional meaning of the word is 'the line where earth and sky meet'. Only at the end of the nineteenth century was the term also used to refer to the outline of buildings on the horizon. Nowadays the words skyline and horizon have to be distinguished: horizon is reserved for the line between land or sea and the sky; while skyline is used for the urban setting of the line between the sky and the city buildings, the urban contours.

Throughout history there have been high buildings, from the *ziggurat* in Mesopotamia to the Eiffel Tower in Paris. Their height was used symbolically. The skyscraper was a new type of building and marked a new direction. This tower was erected for functional purposes, and the symbolism was a bonus (Kostof 1991: 279).

Another difference between high buildings in history and those of today is that the historical edifices had a public function. They were used for religious or governmental purposes, or to proclaim technological advances. Incontrovertibly, the skyscraper is the product of private initiative.

A building from the past comparable with a modern skyscraper, in the sense of its use and the fact that it is private property, is the tower-house of the Middle Ages. Most of these towers were situated in cities in northern and central Italy, in the south of France and in central and southern Germany. Their function was to provide defence as well as to display the affluence of the various clans. The towers in Germany were crude and squat, whereas in Italy they were tall and slender. The Italian tower belonged to a powerful family or a group of families, the so-called tower-associations, which availed themselves of the facilities and the protection that the towers afforded (Kostof 1991: 280–1).

There are two ways to achieve a remarkable skyline. One is to make use of natural landmarks already present. This is the case for instance with Sugar Loaf in Rio de Janeiro and Table Mountain in Cape Town. The other way to create a spectacular skyline is to use architectural expressions such as the Eiffel Tower in Paris and St Peter's Basilica in Rome.

A city skyline is not normally determined by just one or two building styles, but by repeated architectural forms such as minarets, domes, industrial chimneys and the like.

Bearing in mind topography, which forms an important part of the skyline, a distinction has to be made between towns in a flat landscape and towns in mountainous or in hilly terrain. In Holland, for instance, windmills face no natural competition whatsoever from the flat landscape. More topographically challenged, important buildings in hilly or mountainous areas are usually built on the top of a hill. Massa Marittima in Italy is a fine example of a town with a church on a hilltop, as is Sacré Coeur in Paris (Kostof 1991: 288–90).

Before the advent of the secular state, the skyline was formed by religious buildings. In Europe these were the high domes and belfries, or an impact was created by building imposing facades such as in Cologne. After the introduction of the secular state, the most important facet of the skyline was the dome. This architectural shape is found on governmental buildings, public baths, mausoleums, and as before on churches. The belfry too is still a frequently recurring architectural form on the skyline of the secular state. In the past they belonged primarily to churches, but now they are also used for government buildings. Frequently, church and state competed with their towers. Height was obviously one of the elements in this competition, but they also tried to outdo each other through the use of various building materials. The towers of Siena and Florence are good examples of this. Other remarkable building projects of the secular state have been lighthouses, watchtowers and water towers (Kostof 1991: 288–306).

In the modern world the skyline is dominated mainly by skyscrapers. These are monuments that emphasize self-interest as well as aggressive capitalistic competition. Because the skyscraper primarily symbolizes *laissez faire*, it is above all a typical American type of skyline. Around 1870 the first skyscrapers were built in Chicago and New York. They became representative of a purely business-like skyline. It was essential that the architectural character of the public monuments of government and culture be different from those of the business skyline. Public buildings such as Grand Central Station, the Metropolitan Museum of Art and the Public Library in New York City express this distinction (Kostof 1991: 323–5).

City towers

When we talk about remarkable skylines in which towers play a major part, the Italian towns in northern and central Italy are the most splendid examples. The best known is San Gimignano in Tuscany, whose towers announce its presence from a long distance (see Figure 1). Around the thirteenth century more than seventy tall, slender towers were built here by rich and powerful families (Raffa n.d.: 34). The skylines of other towns in Italy, such as Bologna and Genoa, were also distinguished by these family towers.

The family formed one of the most important cornerstones of Italian society in the Middle Ages. Families were large and family members often lived together under one roof in large houses with many floors. Less affluent inhabitants also lived in such spacious, high houses, often more than one family occupying the same building.

Around 1150 the richer inhabitants began to replace their wooden colonnades with stone arcades and they built houses of four to five storeys, using stone up to the third floor and bricks above that. This style of building with two different materials remained a characteristic feature of Genoese domestic architecture throughout the centuries (Heers 1977: 132).

Figure 1 San Gimignano's skyline (Photograph: Jan and Monique van Gent, 2005).

Medieval houses in Pisa, for instance, were quite narrow and were supported by two or more pillars and arcades. Most of these houses were between 10 and 20 metres high. This type of dwelling increasingly began to resemble a tower.

Because this kind of house was found more often than any other type on the Italian peninsula, Italian towns formed a huge contrast to other towns in Europe. In Germany, for instance, tall houses were also favoured, but in a much smaller number and with 'only' three or four storeys (Heers 1977: 131–3).

These tall houses can also be found in other parts of the Italian peninsula, and even across the border, owing to the conquests of the Genoese. In Provence (Grasse and surroundings, Saint Tropez and Pontevès), at several locations in Corsica (Bastia, Bonifacio), but also as part of the distant Genoese trading posts on the Black Sea, we find this type of house. Caffa it seems would have looked very similar to Genoa.

In Italian towns such as Pisa and Lucca (Figure 2) these tall houses remained the fashion for quite a long time, in contrast to Florence. Here, many of these constructions disappeared towards the end of the Middle Ages, because the various clans split up and dispersed (Heers 1977: 134–5).

The question of why these tall, tower-like houses were built leads to many different answers. Protection against the wind is quoted as a reason, but a lack of space is also suggested. The fact that towns are built against hills or in narrow valleys plays a part in their being constructed upwards. Genoa, at the foot of the Apennine Mountains and with the sea on the other side, is a beautiful example of this. But this explanation is too simple for Genoa, since the fact remains that there was enough space for the town to spread farther out onto the coastal plains of Polceverra and Bisagno, quite apart from the fact in those days there was enough room left to build within the city walls.

Figure 2 Multitude of towers in Lucca, fifteenth century (Reprinted from Heers 1977: 187, with permission from Elsevier).

It is more probable that these towers were built for reasons of political and military weakness, and perhaps secondarily as a result of the dearth of surrounding countryside in which defences could be set up. But the most important reason why these houses were built this way was because they had such an immense symbolic power, representing the strong unity of the families, therefore reflecting their riches and power. A tower was a true status symbol (Heers 1977: 134).

In his chapter about palaces, Mumford (1961: 378) refers to the family towers in Italian towns. The word *palazzo* refers not so much to a palace as we know it from fairy tales, but to a large and beautiful building where a rich merchant or some other important gentleman may have lived. The English adjective palatial in baroque terminology stands for spaciousness and self-fulfilling power. Mumford quotes the towers and their families in towns such as Lucca, Bologna and San Gimignano as examples of this self-fulfilling power in the fourteenth century. This form of vertical power was, he argues, the trademark of the fourteenth century. Besides the extended families, there was often also a close relationship with the neighbours. Guglielmi (1994) dedicates a whole chapter to this relationship.

Many towers were given the name of the families to which they belonged, and the towers which were captured during the Crusades were also often given the family name. These towers were the city's pride and the largest towers were therefore often given a nickname as well as the family name. In Lucca the towers which were the pride and joy of the town were called Lancie, Guardamonte and Torre de Levre (Heers 1977: 188).

Noble families frequently inherited ancient Roman monuments, which they used as they were or adapted. In Italy, Hadrian's mausoleum was turned into a medieval fort. Many old buildings were also used as they were or adapted, such as the Miranda Tower of the Arcioni family, which was built on the ruins of the Temple of Jupiter. Later, the old Roman walls were used or rebuilt as a defence, but the clans especially attempted to obtain a tower. These towers were almost never intended for habitation, but served as a place of refuge and symbol of power (Heers 1977: 176).

The tradition of towers harks back to Constantinople in the Middle Ages, where Theodosius' fortress wall boasted ninety-six towers of various sizes, which had been allocated to the owners of the adjoining parcels of land. Around 1420 Rome still had a large number of towers that were in a good state of repair or had been rebuilt: forty-nine in the area between the fortress walls of Porta Appia and Porta Ostiense, and twenty between Porta Metronia and Porta Latina. Here, too, most towers were owned by large families (Heers 1977: 177–8). Even if towers were not an integral part of a house, the homes of the nobility in the Middle Ages still resembled forts. In Italian texts, therefore, a distinction is made between towers and houses with towers, the so-called *case-torri,* which were fortified houses with at least one watchtower each (Heers 1977: 200). A *casa-torri* was usually rectangular, whereas a tower had a square base. The tower was not intended to be a dwelling space, whereas the *case-torri* did function as such, and also served as a strategic point in defence whenever a conflict flared up between two rival factions, as in Bologna with the Lambertazzi and the Geremi (Rivani

1966: 10–11).

Tower associations

Apart from symbolizing the affluence and power of a family or clan, these houses were also built for military purposes. There were tower associations in which the heads of families who owned one of these fortified towers were united. These tower associations played a large part in the social, political and economic life of medieval cities, especially in Tuscany.

The first Florentine tower associations brought a considerable number of people together, principally heads of families. The first treaty was signed in 1178 in which were mentioned the names of the sixteen families involved in the building of two or more towers. In these agreements it was also decreed that *rectores*, who were responsible for overseeing the building of the towers, spread the cost of the building work and solve any possible internal conflicts, be appointed. So the towers were communal property, divided into shares. Every share was to be inherited and in the event of the death of an owner passed to his eldest son. The rule was that the members of the association had to be prepared to assist each other in the event of the involvement in an external conflict of any one member (Heers 1977: 194–5).

Notarial deeds also show that the associations often remained intact for a very long time. This has been concluded from a deed of sale of a quarter of a Florentine tower. In this 1174 deed certain conditions were stipulated, which reflected some traditions, or at least a few old customs, that had been instigated in the past and were still legally valid (Heers 1977: 195).

In Pisa the tower associations of the thirteenth century were often maritime associations, responsible for the vessels of the communal fleet. In the Genoese fleet, too, ships which belonged to a family or clan set sail flying a flag with the family colours, taking their place alongside the vessels of the admiral or the commune, the city state (Heers 1977: 170–1).

In the years 1170–1190 there was a real upsurge in the building of towers throughout the Italian peninsula. In Florence, the Lamberti family erected their huge tower in 1172; the Bevenzini built theirs in 1180. In Siena the two towers of the Gallerani family were constructed in 1186 (Heers 1977: 179). But how many towers were there exactly in every town? And what were the deciding factors which determined how many towers there should be?

Sources

There are various types of sources from which information about the towers, such as the number that existed and their differentiating characteristics, can be delved. The most important sources are documents, stories and illustrations.

In texts a huge interest in the towers can be traced. When it comes to the density of the towers in the Middle Ages, in the past there has been some exaggeration. Travellers and writers of chronicles often happily overestimated the number of towers in a town or city. Despite such hyperbole, that there were many is beyond a doubt, because this is the one fact they all emphasize. The towns which are always mentioned are those that can be seen from a long distance, thanks to their many tall, slender towers.

Davidsohn visited Florence in 1180 and counted 135 towers, as he recounted in his book *Forschungen zur älteren Geschichte von Florenz*. But he thought the true number of towers must have been three times as many (Heers 1977: 188).

In Rome forty-four towers could be counted in a single *borgo*. In some places there was talk of a veritable *selva di torri* (forest of towers) inside a city (Figure 3).

Figure 3 Selva di Torri, Bologna, Middle Ages (Source: Arbizzani 1958: annex).

Gozzadini carried out an extensive investigation into the number of towers in Bologna. For this he consulted many types of documentation in the archives, which enabled him to reconstruct the city landscape of the Middle Ages. Among these documents there were 240 official deeds from after the year 1300, which related to the ownership and use of towers. This refers to a fairly late period, when the families had probably already lost or had been forced to pull down a large number of their towers, as their power was ever-increasingly threatened by the power of the commune.

In spite of this, Gozzadini has been able to prove beyond any doubt the existence of 194 family towers. Of fifty-four of these towers he has described their width and the thickness of the walls. Having done so, he has indicated street by street how many towers it contained: in the Strada Maggiore there were eighteen such noble towers, in the Via Porta Nuova ten, and thirteen stood in the Via Castiglione (Heers 1977: 189). The density of the towers must have been extremely high. So here there was also clearly talk of a *selva di torri*. When it comes to the level of symbolism of the towers, a shift can be observed from 'family' to 'city', because of the large number of towers along the city walls.

Iconographical documents which reproduce the medieval townscape, often on a hill, also emphasize the important and central role of the large number of towers on and within the walls. In the backgrounds of the works of all Tuscan and Lombardian painters from the thirteenth to the fifteenth century, the town is seen as part realistic, part idealized landscape. The town is surrounded by strong walls and tall family towers dominate the skyline, and are often represented as much taller than those on the city walls and even those of the churches (Heers 1977: 182).

These city views also appeared on contemporary seals. They depict stylized, symbolic reproductions of towns reduced to their essential elements: city walls and many towers that rise above these. Hence the strength of the families, or in other cases, of the communes, was shown (Heers 1977: 184).

In art there are two types of representations of the medieval towns and their towers (Heers 1977: 185). There are those in which the Madonna blesses the town and protects it in the folds of her gown. The other views are representations of dedication to the Blessed Virgin, in which a bishop or patron saint is depicted offering the city, which he holds in his hands in miniature form. A beautiful and well-known example of this is the altar-piece of the eponymous patron saint of San Gimignano, who holds the city on his lap, painted by Taddeo di Bartolo (1362–1422) (Figure 4). In this manner many towns were represented: San Gimignano with its strong, tall, slender towers standing apart from each other and Pisa, for example, with its wider towers in compact blocks.

Figure 4 The patron saint of San Gimignano (Italy) with the city crowned by feudal towers on his lap; late fourteenth-century altar-piece by Taddeo di Bartolo (Source: Raffa n.d.).

For a long time towers were thought of in terms of townscape. They often marked the ancient city centre, the spot where the most important clans, who had founded the city, lived and reigned, and in this way it distanced itself from the areas outside the centre where newcomers had settled. This is seen in one of the illustrations of Giovanni Sercambi in the *Cronaca*. It shows how a wall is being built outside the city of Lucca to protect the outlying areas (Figure 5). The towers of the city, however, are situated inside the already circumvallated centre, and in the new area no towers are to be found at all (Heers 1977: 188).

Figure 5 City wall around Lucca. Drawing after Giovanni Sercambi's illustration in the
Cronaca showing how a wall is being constructed outside Lucca to protect the environs
(Reprinted from Heers 1977: 186, 185–8, with permission from Elsevier).

Other features of the towers

Towers did not look the same everywhere. There was a difference in the
materials used, colours, shapes of windows and battlements and, of course,
height. Thanks to quite detailed deeds, it has been possible to describe the
towers of various towns quite accurately.

In Florence, Lupo di Castiglionchio shows how the 150 towers which
were built at the beginning of the thirteenth century must have been 70
metres high; the tallest one was 76 metres.

The towers of San Gimignano did not reach this height, but were equally
impressive and stood on strong foundations (Heers 1977: 191). Here, the
slots in the walls, that were used to be able to build gangways quickly, can
still be clearly observed. These bridges served to move from one tower to
another, but could also be used to attack someone else's tower (Raffa n.d.:
35).

In Siena, the towers were taller and more slender as they were in Lucca.
In northern Italy the towers were also tall and slender and invariably square,
and definitely higher than the houses. The most beautiful example of this is
the Asinella Tower (Bologna), which is 97 metres high.

In Genoa stands the Embriaci Tower, brooding with its walls of dark
stone without one single window, hidden by the houses crowded around it.

From out at sea it is obvious how this tower appears to dominate the town: slender, situated in the middle of the town on the hill of Sarzano, 78 metres high (Heers 1977: 191–2).

Initially the towers served as places of refuge during wars and times of conflict. They were completely devoid of any form of convenience. When they were later used by one family, rather than by several groups of families, some storeys were turned into habitable spaces and the towers were then used on a more permanent basis. This process was mainly carried out between the twelfth and fourteenth centuries, but even later than that the fort-like towers were turned into living accommodation. To make them liveable, enormous changes were made to the exteriors and interiors of the towers (Heers 1977: 192).

On the ground floor the whole of the space would be taken up by one room. This room normally had a stone vault and very thick walls. Its purpose was to be a storeroom for provisions and an arsenal. Later it would be used as a meeting room where relatives and friends could assemble. This space was not normally accessible from the street; the walls generally would not have had any gaps, and if there were any, then these would have had to have been extremely narrow as they had to be easy to defend.

Entrance to the tower was usually gained through one of the upper storeys, by means of ladders, or over suspension bridges from the neighbouring properties to the windows of the tower. Inside the tower there would be a staircase to the lower floor and staircases to the floors above and to the flat roof, fringed with battlements. Throughout the previous centuries, when the towers still had a purely defensive function, stones, canon balls and all sorts of projectiles were hurled in the direction of the enemy from these flat roofs. Archers positioned themselves on the various floors and shot their arrows through the windows or through the narrow observation holes (Heers 1977: 193).

Initially for defensive reasons there were very few gaps in the walls. Those there were, were normally narrow and rectangular; usually there were only two or three on any one side of each tower at great height. Once the towers began to be used more for domestic purposes, more openings in the walls appeared and they were more highly decorated: windows with round arches, much wider than before and often in pairs (Heers 1977: 194).

The tower would be situated among the houses of the clan's *contrada*. Initially, families attempted to build their towers immediately next door to their homes, so that they could pass easily to the tower by means of an interior corridor. This traditional layout of a tower immediately next door to the home is still clearly visible in San Gimignano, where the Torre degli Oti is situated right next door to the Oti Palazzo. Whenever two or more families owned a tower, it was erected at an equal distance from the homes of the members (Heers 1977: 195).

A threat to the towers

Because the towers always symbolized the power of the nobility, they were in themselves a form of resistance against the state and against the transition from a feudal political regime to a more centralized system. This concept explains why there was talk of a systematic destruction of the towers, so-called contra-symbolism, by order of other rulers after an uprising or victory. This happened, for instance, by order of the German emperors in Lombardy. Frederick Barbarossa even destroyed the belfry of the Cathedral of Milan.

In Central Italy the towers were more likely to be threatened by the activities of the commune. From contemporary chronicles we learn that destroying the family towers was one of the more important activities the communes undertook in protest against the political and military powers of the clans. This process of demolition started early, but enjoyed only intermittent success. Therefore the systematic demolition of the towers never eventuated (Heers 1977: 197).

Often, the nobility also formed part of the commune and consequently guaranteed cohesion between the tower associations and so attempted to protect these. Transfers of ownership, which would have weakened the group, were forbidden or, at least, discouraged. In this way the cohesion between the various groups was upheld for many years; during years of war and conflict it was made even stronger. For this reason there was no chance of a systematic demolition of the towers. Shares in the towers could only be sold if the owners did not have a male successor, and even then permission from the *rectores* was needed (Heers 1977: 198; Kostof 1991: 280–1).

In the beginning, the communes only wanted to limit the number of towers, to keep a grip on power and on the use of the towers. Pertinently, towers which belonged to rebels or banished individuals had to be destroyed. In Rome this had already been done in 1114.

The commune also drew up rules to limit the height of the towers, as in Genoa in 1143. Only the Embriaci family was allowed to keep their tall tower, which was much higher than any other in the town; this was a reward for their services in the Crusades to the Holy Land (Heers 1977: 198).

In the fourteenth century in Florence rules were drawn up to stipulate that all towers which were in a bad state of repair and posed a danger to their surroundings had to be demolished. At the same time a prohibition was placed on building towers over 30 metres high and on putting catapults on the top. The limit on height which was stipulated was a blow to family pride and to the symbolic power of their tower. Fierce resistance to these rules arose. Consequently, only a few towers in Florence were either lowered or demolished. It was the same story in other Italian cities. At the end of the day it was often easier for the communes to tackle the direct political power of the clans than to destroy their military prowess or external signs of pride and power (Heers 1977: 199).

It is more likely that the towers disappeared from the townscapes simply because of a lack of finance, as a result of which they were neglected and deserted. Improvements in defence and fighting techniques, and the higher standard of living, led to the towers losing their function to defend and protect. The townscape changed as a result of this, but only very slowly and barely perceptibly. Towards the end of the Middle Ages there had been great changes in the towns of Italy, but most of these settlements still possessed their crown of towers (Heers 1977: 199–200, 1984: 160).

In towns such as San Gimignano and Bologna these family towers can, as indicated earlier, still be visited. But in other cities there are also still traces of the *selva di torri* to be found. Even if the towers have shrunk to the height of the neighbouring house, the old features of the nobles' towers are still clearly recognizable. The height of what has in the meantime become a house or home, the narrow gable with usually only two windows on each floor, and most importantly of all, the staircases on the side which allowed entry to each floor, are features reminding us of the fortified towers with their ladders on the outside or their interior staircases. All this is proof of a preference still to protect and isolate the living accommodation (Heers 1977: 201).

The Asinella

In order to get a better picture of the towers, I shall take a close look at both the Asinella and Garisenda Towers in Bologna. The Asinella on the Piazza Ravegnana is the tallest in the city, soaring 97.2 metres. Immediately alongside stands the Garisenda Tower, with a height of 48.16 metres. This tower must have once been around 70 metres high, but in the fourteenth century it was partly demolished, because it had developed a considerable list and was becoming a danger to its surroundings. For this reason it earned the nickname La Mozza (The Chopped One). At the moment the Garisenda is still listing as much as 3.22 metres. The Asinella is not leaning quite as much, only 2.23 metres. From the top of the 498 steps that form the staircases in this tower there is a beautiful panorama of Bologna and its environs; when the weather is clear it is possible to see the Adriatic (Arbizzani 1958: 50). In the past there must have been 200 towers in Bologna; now there are only a few left. The slender, delicate Asinella was built between 1109 and 1119 by Gherardo Asinella, and the Garisenda in 1110 by Filippo and Addo Garisendi (Arbizzani 1958: 50).

The Asinella is most renowned for its unusually thick walls, which at the base are between 3.75 and 4.1 metres thick. The Galluzzi Tower in Bologna is the only one that comes anywhere near the Asinella, with walls of 3.1 metres (Rivani 1966: 13). The base of the Asinella measures 8.78 on the north and south sides, 8.8 metres on the west side and 8.74 on the east side (Rivani 1966: 14).

The door on the ground floor is in the southern side of the tower. A second door is situated in the eastern side and used to lead to the first gallery. Above this base the tower thrusts into the sky in two connecting parts shaped like the base of a pyramid. The first part reaches a height of 34.4 metres, and where it transforms into the second part, it is finished off with battlements and has a robust crossed vault as its ceiling (Rivani 1966: 16–8).

In the first part above the base there are two windows, one on the front of the building at a height of approximately 17.65 metres, with a remarkable brick bow on the inside. This may have been the entrance to a third gallery, because at around 5 metres above the first gallery are the remains of consoles which would indicate that there may have been a second gallery there. This would have been accessible by means of wooden ladders which would have leaned against the outside wall of the building. The next window is situated at a height of 25 metres on the southern side of the tower. A third window is situated near the vault on the northern side of the tower between the battlements which protrude around 30 centimetres at the base of the second part of the tower.

The thickness of the walls at this height, at the end of the second part, is around 2.1 metres. The width of the walls of this part is 8 metres at the base, but only 7.55 metres at the battlements (Rivani 1966: 18–20).

The sides of the second part of the tower are 6.95 metres wide at the base and above, where the battlements are situated, only 6.5 metres. In this part of the tower, at a height of 44 metres, there is a window on the eastern side and another one on the southern side at 48 metres. A further window is situated on the western side at the point where the second floor commences above the arches, at approximately 57 metres. On the third floor, where the transmitters of RAI TV are housed, another window is situated on the northern side. Here, the thickness of the wall is only 1.15 metres. The fifth window is situated at a height of 80 metres, where the wall is only 1.05 metres thick. The last window is situated on the northern wall, on the fourth floor at 85 metres high, immediately below the balcony which indicates the top of the tower. The walls are only 0.96 metres thick here (Rivani 1966: 20).

In 1185 the Asinella Tower fell into enemy hands and was set on fire. The interior was completely destroyed. In 1286 the Asinelli regained possession of the tower and that same year the commune commenced planning the Piazza Ravegnana (Rivani 1966: 26). To construct this square they demolished many properties, among them the house of the Asinelli, in order to isolate the two towers: '*ridurre in isola le due famose torri Asinella e Garisenda*' (Heers 1984: 157n). This work was completed in 1288 and subsequently the commune bought another three parts of the Asinelli tower. In 1292 it leased a further part and in 1306 it purchased yet another section of the tower.

The following years were turbulent. A fire destroyed the wooden staircase on the inside, but the walls were left fairly undamaged. In 1398 the tower passed completely into the hands of the commune. In 1413 another fire was lit in the tower (Rivani 1966: 26). Later a clock was placed on top, which was given the nickname *del fuoco*, because it warned the city every time a fire broke out. On the side of the Strada Maggiore, an iron cage was later hung outside a window to house criminals (Arbizzani 1958: 50, Rivani 1966: 26–8).

After having fallen victim to various arsonists throughout the ages, the Arsinella has been regularly affected by earthquakes. But the tower has resisted these valiantly. The tremors of 1399 and 1505 were the most serious, but even then only the dome and one battlement collapsed. Its greatest enemy has been lighting, which has done more damage than any fires or earthquakes (Rivani 1966: 28).

The Garisenda

The Garisenda is situated only eleven metres from the Asinella. Together they form the remarkable skyline of Bologna. It has already been said that this tower is known mainly for the fact that it is listing 3.22 metres in a southeasterly direction. The ground on which the tower stands is the reason that it is sinking. Even in the days of the poet Dante Alighieri (who knew the tower when it had its original height of 60 metres) the tower was already leaning badly. The Garisenda is mentioned by Dante in his *Divina Commedia*, in the XXXI verse of *L'Inferno* (Alighieri 1313–1321). This text is also found at the foot of the tower on a tile in the wall (Rivani 1966: 33):

'Qual pare a riguardar la Garisenda
sotto il chinato, quando un nuvol vada
sovr'essa sì, ch'ella in contrario penda;
tal parve Anteo e me che stava a bada
di vederlo chinare, ... '[1]

The Garisenda has a base of approximately 8 metres per side and the thickness of the wall at the base measures 2.35 metres. The door on the ground floor is on the western side. Halfway through the fourteenth century the tower was drastically shortened and now only reaches a height of 48.16 metres. The walls at the top are 1.95 metres thick: a proportionally much smaller decrease in wall-thickness than is the case with the Asinella.

A second door is situated on the northern side at around 6 metres above ground level. This initially connected to the top floors of the Garisendi houses by means of galleries and wooden gangways. On the other side of the tower is a window of 60 centimetres wide and 1.80 metres high. Each of these windows was originally connected to a gallery which was supported by the characteristic consoles of undressed limestone, of which traces are still visible (Rivani 1966: 34).

Just like the houses of the Asinelli, the houses of the Garisendi around the towers were also demolished in the thirteenth century, in order to construct the square which the commune had in mind. Between 1351 and 1360 the city fathers decided that the tower had to be shortened, because the listing of the building constantly increased the danger of collapse. The tower ended up at its present height (Rivani 1966: 41).

This leaning of the tower greatly inspired the fantasies of previous generations of Italians. There is a story that a Garisendi, who was looking for a wife, let his eye roam over a young Asinelli lady. When he asked for her hand, he got the answer that the person who could build a tower equally as beautiful alongside the Asinelli tower, would win the lady in question. The Garisendi man consequently built the tower in such a manner that it seemed the two towers almost touched each other (Figure 6). This way it almost

---

[1] As the Garisenda looks from underneath
its leaning side, at the moment when a cloud
comes drifting over against the tower's slant,
just so the bending giant Antaeus seemed
as I looked up, expecting him to topple ... (Alighieri, translated by Mark Musa, 1971).

appeared as if the two towers were kissing (Rivani 1966: 44).

Figure 6 The Garisenda Tower apparently kissing the Asinella Tower (Source: www.norreg.dk/ geografi/98prog.htm, accessed 15 September 2004).

Conclusion

Throughout the ages tall towers have always been one of the most important features of the skyline. Even now the skyscraper is still one of the ultimate ways to symbolize the power and wealth of a city. The resemblance, for instance, between the (former) skyline of Manhattan and that of San

Gimignano is striking (compare Figures 7, 8 and 9). Therefore, it is interesting to draw a parallel between the towers of earlier centuries and their modern counterparts.

The medieval towers in Italy belonged to powerful, wealthy families. By building these towers they displayed their prestige and wealth. Besides being a status symbol, the family towers were also built for purposes of defence. In cases of emergency, the families or clans could shelter inside the tower. The towers then – in the case of the Ardinghelli even twin towers (see Figure 9) – were clearly a personal symbol.

Threats to the family towers were posed by rival families, town officials or other rulers, but also from the communes. These wanted to put an end to the feudal system within the city walls and aimed to achieve a more centralized political system. It was necessary therefore to erase the symbols of the feudal rulers: the towers had to come down. This policy was not always successful.

It seems that almost nothing has changed today. Towers are still being built to function as status symbols and to display wealth and prosperity to the outside world. Even today there is a competition as to which city has the tallest towers. Although many family towers still exist, for instance, the Rockefeller Center in New York City, most skyscrapers are being built for economic reasons. From being personal symbols, they have become public symbols. On the relatively small island of Manhattan people had to build into the air. The towers no longer serve to defend as in days gone by. The most important use of skyscrapers is to house offices and towers have become the trademark of the business side of the city.

Towers still retain a certain symbolic function. They blatantly represent the world of capitalism. This became clear by the attacks on the World Trade Center in New York City of 11 September 2001. If a blow was to be struck at the heart of the capitalistic world, then the Twin Towers were the unmistakable target. And so it came to pass …

After 9/11 the symbolic saturation with regard to the city towers has ebbed enormously (Sluis 2003: 258–9). Before then, people were not really aware of the symbolic power and meaning of certain areas or buildings in cities. After the attacks on the heart of the capitalistic world, every inhabitant of and visitor to such an area or building has become more conscious of the fact that such an architectural expression is the very symbol-carrier of certain features of a society, which makes it into a possible target for acts of terrorism aimed at all that this society stands for.

Figure 7 The former Twin Towers as a central element in the skyline of Manhattan
(Photograph: E.D.M. Both, 2001).

Figure 8 San Gimignano's towers (Photograph: Jan and Monique van Gent, 2000).

Many a tourist will stop to think before going to the top of the Empire
State Building. So even today towers still often represent the symbolic core
of power, and moreover they are also being destroyed by enemies, because a
direct attack on real power is more difficult, if not impossible.

Figure 9 The Ardinghelli twin towers in San Gimignano (Source: Raffa n.d.: 20).

References

Alighieri, D. (1313–1321) *Divina Commedia.* [Translated by Mark Musa, 1971, *Divine Comedy*, Indiana: University Press]

Arbizzani, L. (1958) *Torri e Castelli: Bologna e la sua Provincia.* Milano: Legatoria Fraschini.

Guglielmi, N. (1994) Vecindad y solidaridad en la Italia medieval, ss. XIV–XV. In: M-Th. Caron, *et.al.* (1994) *Villes et Sociétés Urbaines au Moyen Age: Hommage à M. le Professeur Jacques Heers*, pp. 259–67. Paris: Presses de l'Université de Paris-Sorbonne.

Heers, J. (1977) *Family Clans in the Middle Ages: A Study of Political and Social Structures in Urban Areas.* Amsterdam, Oxford and New York: North-Holland Publishing.

—— (1984) *Espaces Publics, Espaces Privés dans la Ville: Le Liber Terminorum de Bologne (1294).* Paris: Editions du Centre National de la Recherche Scientifique.

Kostof, S. (1991) *The City Shaped: Urban Patterns and Meanings through History.* London: Thames and Hudson.

Mumford, L. (1961) *The City in History: Its Origins, its Transformations and its Prospects.* London: Secker and Warburg.

Raffa, E. (n.d.) *San Gimignano by the Beautiful Towers: Guide-Book Art and History.* San Gimignano: Brunello Granelli's Edition.

Rivani, G. (1966) *Le torri di Bologna.* Bologna: Tamari Editori.

Sluis, R.J. (2003) 'A reassessment of Lynch's conception of meaning.' In: P.J.M. Nas, G. Persoon and R. Jaffe (eds.) *Framing Indonesian Realities: Essays in Symbolic Anthropology in Honour of Reimar Schefold*, pp. 251–74. Leiden: KITLV Press.

Chapter 4

LOOKING BACK, FORWARD OR UP?
Urban symbolism and the politics of race in Jamaica's Redemption Song

Rivke Jaffe

This contribution aims to give a short exploration of the symbolic order of the city of Kingston, Jamaica. By examining a specific urban marker – the sculpture 'Redemption Song', intended as the highlight of the city's Emancipation Park – I will investigate the contemporary discourse of slavery, 'race' and colour and the dynamics of contested space in a post-colonial society. The symbol's significance is examined in the context of the construction of national identity, a process that is intimately linked to local power relations and ethnic discourse. This study of the symbolic domain is based on a review of leading Jamaican newspapers centring on the debate surrounding the sculpture which took place in August 2003 and following months.[1]

Following a brief historical overview of Jamaica and more specifically Kingston, the sculpture as well as its location will be discussed. Next, a closer look is taken at both the production of this symbol (what it was intended to signify and express) and its consumption (how its meaning was perceived and constructed by the larger public). The discrepancy between these two processes led to a controversy surrounding the sculpture which must be seen in the light of Jamaica's history of slavery and institutionalized

---

[1] In addition, participatory observation and a series of twenty mini-interviews were conducted in the park in March 2004. However, the material collected through these methods corroborated rather than supplemented the other data so the focus here is mainly on newspaper articles, editorials, columns and letters to the editor in the *Jamaica Gleaner* and the *Jamaica Observer*.

racism resulting in the contemporary politics of race.

Kingston, Jamaica

Jamaica is the largest of the Anglophone Caribbean islands. It was captured from the Spanish in 1655 by the British, who proceeded to colonize the territory and grew wealthy developing sugarcane plantations, making extensive use of slave labour. The white elite was always very small; Europeans never represented more than a small percentage of the population. A so-called plantation society developed, characterized by economic dependency and 'an abiding Eurocentrism which puts everything European in a place of eminence and things of ... African origin in a lesser place (Nettleford 1978: 3). In 1807, Britain declared the slave trade illegal, but slavery itself was not abolished in the British colonies till 1 August 1834. This was followed by a period of what was euphemistically called 'apprenticeship' in which ex-slaves were forced to continue working for their former masters, supposedly to gradually accustom all involved to freedom but actually to prolong the supply of free plantation labour. Due to a marked lack of success, this system was brought to an end prematurely, on 1 August 1838.[2] Most emancipated slaves left the plantations as soon as possible, which resulted in a large labour shortage for the plantation owners, mitigated somewhat by the advent of indentured servants from India and China. In 1962, Jamaica was granted independence from Britain. These three institutions – slavery, the plantation system and colonialism – are the historical context from which the social, cultural, political and economic patterns of modern Jamaica can be said to derive (cf. Randall 2003).

Kingston was founded in 1692 on the coastal edge of the Liguanea Plain, after a large earthquake destroyed the city of Port Royal, which was located on the opposite side of the harbour on the Palisadoes peninsula. The colonial capital at the time was Spanish Town; Kingston's principal role was the collection, storage and export of sugar to Britain, and the import, storage and dispatching of slaves and British manufactured goods to the plantation estates. Fortunate to have an excellent natural harbour, the city became the trade capital of the region, a focal point of the 'triangular trade' between Britain, Africa and the New World as well as regional trade. The population grew due to (involuntary) immigration from Africa and Europe and in-migration from the plantations. Kingston's social order was liberal compared to the estates, but still very much related to colour, with a strong colour-

---

[2] It is for this reason that some consider 1834 the year in which Emancipation occurred and others 1838. In both cases Emancipation Day is 1 August.

status correlation. Spatial segregation was never very rigid, but as Clarke (1985: 162) states, 'where status is highly ascriptive, space is often unnecessary as a social barrier'. As the city expanded, richer people chose to live in the cooler, less dusty hills, so that there is a certain relationship between income and spatial distribution. Up to the present day, neighbourhoods in the southwestern parts of the city are still visibly poorer than those to the northeast (Clarke 1975, 1985).

After emancipation and the onset of the free trade era, Jamaica and Kingston's economies slumped. Kingston's role as a trade centre diminished and the harbour's commercial activities grew less. With the exception of a small economic comeback when Kingston became the capital in 1872 and a short banana export boom in the 1890s, this economic recession lasted until the early twentieth century. Racial and cultural pluralism became more established in this period, with the three main strata being 'blacks', 'coloureds' and 'whites'.

The 1920s and 1930s were decades of civil and labour unrest as well as Black Nationalist movements such as Marcus Garvey's UNIA (Universal Negro Improvement Association). These awareness-raising factions played a significant role in attaining universal suffrage and later independence, and can be seen as the predecessors of the Rastafari, a socio-religious movement with a strong focus on racial consciousness and repatriation to Africa. In the 1940s and 1950s, Kingston's population once again grew explosively due to urbanization, but the majority of rural migrants found themselves impoverished and living in Kingston's infamous tenement yards. Social conditions were and stayed miserable, with high rates of unemployment, violent crime and overcrowding. Post-independence 'garrison politics'[3] and gang warfare between supporters of the two major political parties did not improve matters, as Kingston's violent reputation scared off cruise ships and tourists. The recent development of New Kingston as the city's modern central business district was more of a success than attempts to redevelop downtown Kingston, but a substantial proportion of Kingston's neighbourhoods are still destitute and crime-ridden. Since the beginning of Kingston's development, a disproportionate number of earthquakes, fires and hurricanes have tormented and partially destroyed the city; this accounts for the low prevalence of historical architecture (Bryan 2000, Clarke 1985).

---

[3] The Jamaican Labour Party (JLP) as well as the People's National Party (PNP) have (had) relationships with their constituency termed 'garrison politics' or 'political tribalism'; in its most basic form, this type of patronage entails that certain neighborhoods are either JLP or PNP dominated and 'only eat when dem party rule' (Moser and Holland 1997: 8–22).

Approximately half of the Jamaican population of 2.6 million resides in urban areas, mainly in the Kingston Metropolitan Area (KMA). Urbanization in recent decades has taken its toll on the natural environment and urban resources. Housing conditions are often miserable, with increasing numbers of people squatting on unsuitable or marginal lands such as steep slopes and gullies. Water supplies, sewerage and waste disposal systems often lack the capacity to provide adequate services to the growing communities. Besides, 'severe stress is placed on ... such factors as quiet, open space, trees, and natural beauty, which are so important to human health and quality of life' (NRCA 1998: 6–7).

Emancipation Park

It is in the context of the urban blight that characterizes much of Kingston that one can view Emancipation Park (Figure 1). This park is a large, clean green space of some six acres in the middle of New Kingston, featuring a large bandstand, a 500-meter jogging track, three fountains, and a variety of mainly indigenous trees, shrubs and flowers. The construction of the park, which was opened on 31 July 2002, the eve of Emancipation Day, was quite costly and its maintenance – security, office, gardening, and cleaning – will continue to be a considerable government expense. Critics saw the park as an extravagance in the light of the country's floundering economy and felt that government expenditures demonstrated skewed priorities. In addition, some felt its location was a waste of profitable real estate. However, as Prime Minister P.J. Patterson (2002) proclaimed in his opening address, the park is a commemoration of the end of slavery, and aspires to 'celebrate and mark a glorious and challenging beginning which had beckoned our forebears to re-discovery of self and the shaping of a new order of society'. In addition, the park's construction 'speaks to such achievements that have come from the attributes that reside in our people' and the PM 'firmly believe[s] that our ancestors would approve of much that has been accomplished'. Moreover, according to the PM, the park is 'vivid testimony to the value we place on urban revitalization', though in this respect one might wonder why the park was constructed in an uptown area rather than downtown Kingston which seems to be in more dire need of revitalization.

Redemption Song

While Emancipation Park by itself – apart from its costs – did not bring about much controversy, the sculpture erected by one of its entrances, Redemption Song, was the cause of quite some scandal. In search of a

Figure 1 Posing for wedding pictures in Emancipation Park (Photograph: Rivke Jaffe, 2004).

crowning artwork for the park, the National Housing Trust[4] together with the National Gallery of Jamaica arranged a competition, which was won by Jamaican artist Laura Facey-Cooper. The bronze sculpture shows a man and a woman, both nudes, standing over three meters high, faces tilted towards the sky (Figure 2). They stand in a pool of water that flows over the domed base of the stature, into which are carved the words 'none but ourselves can free our mind'. This quote by Black Nationalist Marcus Garvey was immortalized by Bob Marley's 'Redemption Song', inspiring the title of the artwork.

The sculpture was revealed on 31 July 2003, a year after the park's opening. Within days the controversy surrounding it had become full-blown; newspapers columnists devoted their space to the topic, readers wrote numerous letters to editors and the various radio talk-shows had hosts, guests and callers discussing the pros and cons of the statues and their location. The main thrust of the debate was the appropriateness of statues' message and location.

Production of a symbol

'Redemption Song' was intended to be a national symbol which could

---

[4] The NHT was responsible for the development of the park.

Figure 2 Redemption Song (Photograph: Claudia Hucke, 2005).

'contribute to sustaining cultural unity and assist in the development of the institutional and operational frameworks for civil society' (Patterson 2002). The sculptor herself elucidated her intent in a press release:

> Even though, nine generations ago, one of my forebears was freed from slavery, it is only recently that I myself have experienced the inner freedom which gives life and meaning to the words first used by Marcus Garvey, then Bob Marley – 'none but ourselves can free our mind'. It was the experience of this message that helped me create the sculpture, Redemption Song. My piece is not about ropes, chains or torture; I have gone beyond that. I wanted to create a sculpture that communicates transcendence, reverence, strength and unity through our procreators – man and woman – all of which comes when the mind is free. To understand freedom, each of us must take a journey within ourselves and make a connection with the divine. … The water is an important part of the monument. It is refreshing, purifying and symbolically washes away

the pain and suffering of the past.[5]

So taking the perspective of the 'production of meaning', the sculpture was supposed to symbolize the actual event, Emancipation, but a stronger emphasis was placed on emancipation as the ongoing process of achieving social and mental freedom. As Sheperd (2000: 58) notes, 'the annual rhetoric which accompanies emancipation day commemoration is very often located within the discourse of slavery rather than within the discourse of anti-slavery and freedom'. Yet, the producers – the artist and indirectly the Jamaican government as initiator – put in place a marker that, in both a literal and a symbolic sense, looked forward and up to the future rather than back to the past. The intended emphasis was on hope and beauty rather than dread and anger.

Consumption of a symbol
The controversy surrounding the sculpture resulted from a discrepancy between the producers' intent and the way in which the symbol was interpreted and its meaning 'consumed'.

Criticism of the statues centred around three main issues. First of all, many Jamaicans objected to their nudity. This was seen as indecent – especially given the fact that many children visit the park for educational and recreational purposes – and gratuitous given the theme of emancipation from slavery. As one park visitor remarked, 'It nuh look good. Dem shoulda cover them up with a loin cloth or something. What the artist is saying about black people? And why dem coulden free up them mind inna dem clothes?' To another, it brought to mind slavery rather than freedom: 'It reminds me of my foreparents, naked in some canefield, and being whipped by Backra master'.[6] Specifically, critics focused on the male statue's genitals, which were seen as being offensively large and obscene. A letter to the Gleaner found it 'in poor taste to have a sculpture with male and female genitals exposed and so exaggerated, erected in a public place ... we need to consider the numerous children who visit the park daily'.[7]

This vehement censure came as a surprise to some, given the overt sexuality of certain aspects of Jamaican culture, such as dancehall music and life-style. It also invited counteraccusations of hypocrisy (see Figure 3) and

---

[5] Source: Jamaica Information Service,
http://www.jis.gov.jm/special_sections/Independence/redemptionSong. html.
[6] '"Renude" controversy', Claude Mills, *Jamaica Gleaner*, 2 August 2003.
[7] '"Redemption Song", bigger but not better', *Jamaica Gleaner*, 5 August 2003.

left a Gleaner editorial wondering whether the criticism 'disguise[d] a deep, unconscious guilt in the psyche of those who express shock at the sensuality of the figures rather than delight in the symbolism the artist envisioned'.[8] However, the explicit sexuality as witnessed in dancehall culture is not necessarily antagonistic to the puritanical sexual norms of 'mainstream' Jamaican society. Rather, the two domains can be seen as complementary and involved in a dynamic dialectic. Strict norms of (sexual) propriety and (Christian) morality – rooted in Victorian England and reinforced by American religious moralism propagated by popular charismatic churches (cf. Austin-Broos 1997) – are inverted in the dancehall. This process is analogous to other ritual inversions and role reversals evident in, for instance, carnival; the dancehall is a bounded, localized and temporalized space with its own strict logic and roles (cf. Cooper and Donnell 2004; Hope 2004; Stewart 2002; Stolzoff 2000) in which definitions of decency and indecency are reversed and contested. However, the inversion of roles and stereotypes of 'decency' outside of this type of established domain and discourse evokes strong negative emotions. Hence objections to the statue explicitly criticize its physical and social location *in a public place*.[9]

A second point of criticism was the artist's background. In the Jamaican context, Laura Facey-Cooper is what is called 'functionally' white, that is somatically and socially she is part of the light-skinned elite. Though she could claim 'at least one' African forebear, by many this was not considered an appropriate 'racial' background for an artist designing a sculpture supposed to represent emancipation – the end of 'white' enslavement of 'blacks'.

After over four decades of independence, Jamaica is after all a country, where, as throughout the Caribbean, a small white minority still exercises 'a political and economic influence disproportionate to their numbers ... shap[ing] the social and economic framework within which the lower classes operate and influenc[ing] policy-making despite the advent of black democracies' (Johnson 1998: x). Bearing this in mind it should come as no surprise that many took offence at a 'white' artist essentially defining freedom from slavery. On the other hand, as the editorial of the *Jamaica Observer* reads, 'we ... have to be careful about expropriating the rights, and worth, of others because their melanin count may not reach, for purposes of populist discourse, a critical threshold'.[10]

---

[8] 'The eye of the beholder', *Jamaica Gleaner*, 6 August 2003.
[9] Many thanks to Douwe Dijksterhuis for sharing thoughts on this topic.
[10] *Jamaica Observer*, 8 August 2003.

Figure 3 Cartoon (Source: http://www.afflictedyard.com/discus/).

A third issue concerned the 'racial' features of the figures. The male statue's large penis was seen by some as a stereotypical representation of the black male, reinforcing the sexualization of the black men, especially given the artist's colour and gender. However, as one letter to the editor of the *Jamaica Observer* stated, 'I bet if they were to reduce the size of his genitalia, a new sequence of controversy would erupt. This time it would be about doing injustice to the proud black man.'[11] On the other hand, *Gleaner* columnist Melville Cooke (2003) lauded the female statue for being representative of Jamaican women and celebrating the black female:

I have looked at the female figure over and over and I have decided to crown her my Miss Jamaica. She is certainly more representative of a Jamaican woman than the often emaciated creatures, always of a, shall we say, light hue who have misrepresented the country with various titles over the years. ...

Let us start with the obvious about the female figure in Redemption Song ... She has the figure of a mature black woman ... Comparing her to any young lady who has held a Miss Jamaica Universe or World title, [the female statue] is more representative of the majority of women I

---

[11] *Jamaica Observer*, 13 September 2003.

have seen in this country – and I have seen a few. When I see a picture of a Miss Jamaica, I see an image that is not only not representative of the majority of Jamaican women, but above that a statement to them that they are not good looking. That to hit the mark they may need to bleach a bit, weave on some hair and get rid of that weight.

The skin colour is obvious – if a monument can be said to have a complexion – and I am ecstatic that there is no long, flowing, burnt out or horse hair sticking from her skull.

To have a prominent artwork apparently celebrating rather than condemning stereotypical 'black' features is indeed a contrast to the barrage of media and other images to which Jamaicans are subjected on a daily basis, promulgating whiteness as the (beauty) ideal.[12] Significant numbers of the women, and men as well, use bleaching creams to whiten their skins, and straightened hair is a very common sight. 'Caucasian' features including light skin, straight hair, thin lips and 'straight' noses are perceived as facilitating social and economic success, and in this bleaching is a rational strategy (Barnes 1997; Charles 2003; Pruitt and LaFont 1995: 426; Sheperd 2000: 54–5).

Another *Gleaner* letter linked the controversy to such feelings of racial inferiority: 'Europe is scattered with statues of nude figures. Why is it that the black man/black woman in Jamaica is still unable to accept themselves as beautiful and celebrate this beauty? Is it any wonder that the majority of black girls in Jamaica apply relaxer creams in order to straighten their hair? Is it any wonder that bleaching creams are so widely used in Jamaica? Blackness in Jamaica is still viewed as inferior and this is a shame. The negative statements emerging about the statue ... reveals that mentally and socially we are still an enslaved people.'[13]

Contextualizing the controversy

The three main points of debate centred on indecency, the artist's suitability, and stereotypical representations of Afro-Jamaicans. All issues were, not surprisingly given the emancipation theme, linked to 'race' issues. These issues were in turn associated directly with current injustices: the continued political, economic and cultural domination of light-skinned Jamaicans, as well as the sustained and renewed world hegemony of Europe and the

---

[12] Television shows, advertisements, video clips and beauty pageants almost exclusively feature and promote light-skinned persons as the somatic norm image.

[13] 'Nude, but not rude', *Jamaica Gleaner*, 6 August 2003.

United States.

These issues are also illustrative of the process of nation-building (in Jamaica and elsewhere): a fairly conscious post-Independence choice was made in the late twentieth century to link national identity to ethnic, or in this case 'racial', discourse. Especially in recent decades, the government has leaned towards 'implicit invocations of racialized frameworks for belonging and ownership' (Thomas 2002: 39). Unlike countries such as for instance Brazil, the US or the Dominican Republic which promote a multi-ethnic or Creole identity, in Jamaica black nationalism and the politics of race have been employed strategically towards the forging of national unity, based implicitly on the premise of ethnic homogeneity (Afro-Jamaicans as insiders) and a common ethnically distinct 'enemy' or 'oppressor' (white European and American outsiders). Despite the presence of some Creole nationalists and the national motto 'Out of many, one people', black cultural nationalism took precedence and Jamaica largely sees itself as a black country. At the same time, the debate surrounding the sculpture highlighted national and international power relations in the construction of national identity: who gets to decide who and what is 'Jamaican'? Who gets to decide what Emancipation looks like to Jamaicans? The fact that a explicitly national symbol, referring to a crucial event in Jamaican history, was created by an artist who according to local ethnic discourse is not a true Jamaican, flies in the face of national identity constructed on the basis of 'blackness'.

The sculpture and the ensuing debate were perceived in a context of Western or 'white' cultural domination, of reflecting external morals and tradition rather than local, Jamaican – interestingly often defined as Christian – values.

Many of those defending the statues referred to the tradition of nude sculpture in Europe and specifically Greece. This justification of the sculpture's integrity in the context of centuries of nudity in European art put some in mind of what Nettleford (1973: 46) calls attitudes of 'civilised "whitedom" over primitive "blackdom", result[ing] from the firmly rooted attitudes of a plantation society'. Another *Gleaner* letter pointed at this attitude and at elite arrogance as follows: 'This is Jamaica so we should adapt the art to the culture and not try to bend the culture to the art. In this time when we speak so much of Emancipation we should seek to rid ourselves of the other side of the plantation complex which says: "We are in the great house and we know what is best for you. You after all are just ordinary Jamaicans and we are the refined ones who can appreciate the fine work of art that we have given to you".'[14] Another concurred with this view, adding that 'it is rather unfortunate that those from the intellectual elite, who would seek to berate [those who object to the statues' nudity] tend to forget that the Crusades, the Inquisition, Slavery, the Holocaust and two world wars were also birthed from the same womb as that of Europe's naked enlightenment'.[15]

Traditionally, the Jamaican and other upper classes throughout the Caribbean have been criticized for being what V.S. Naipaul calls 'mimic men', copying European or North American forms and disregarding local originators. The fact that the statues attracted a 'critique against latent Eurocentrism in the collective mind of the Jamaican populace'[16] is in a way a good sign: referring to European art as the norm and as justification is no longer accepted uncritically by the public nor by (parts of) the intelligentsia.

However, an issue that did not arise – as far as I am aware– is the fact that official monuments and specifically statues are a European or Western tradition that in itself is not rooted in Caribbean culture. Memorializing the past is done in other ways. As Martinican novelist Patrick Chamoiseau writes, 'The Creole city which has so few monuments, becomes itself by virtue of the care it puts into the places of its memory' (quoted in Wheeler 1996: 174). Formal, 'top-down' commemoration as expressed through official symbols is not necessarily the only, or the best, method and Caribbean people have created their own memorials without and outside of official forms.

Caribbean tradition often uses less static, un-localized modes to commemorate the past. The horrors of slavery are kept alive through music, in Jamaica for instance in Burning Spear's 'Slavery Days', Gregory Isaacs' 'Slave Master' and of course Bob Marley's 'Redemption Song'. In Haiti the mayor of Port-au-Prince commissioned several artists to work together on a commemorative mural on a cemetery wall, portraying Haiti's past from slavery to the Duvaliers' dictatorship to current suffering.[17] The mural as a monument is more 'indigenous' to the Caribbean and Jamaica, as witnessed by the countless murals honouring dead heroes (local or national) in Kingston's ghettos. In addition, the fact that the Haitian artwork was a collective effort, and, importantly, that the process of its creation was public, with people passing by continually to observe and comment on its

---

[14] 'Respect each other's opinion', *Jamaica Gleaner*, 7 August 2003.
[15] 'Socialisation not poverty causes crime', *Jamaica Gleaner*, 11 November 2003.
[16] 'Conceptions of Morality', *Jamaica Gleaner*, 8 August 2003.
[17] *Mijn Vriend de Burgemeester*, documentary film, Hans Fels, 2004.

development, made it much more acceptable to the general public, even though it portrayed sensitive themes.

An artwork supposed to symbolize such a sensitive topic as Emancipation could hardly avoid being the focus of much debate; the likelihood of any sculpture being accepted and praised unhesitatingly was fairly unlikely.[18] As the director of a Kingston art gallery put it, 'we all have in our heads our own desires of how we would project in an artistic way our idea of an event as important as Emancipation'.[19] Mainly outside Caribbean societies, but within them as well, the view is sometimes held that slavery 'happened' so far in the past that is does not necessitate commemoration. This does not take into account the degree to which 'sentiments surrounding "emancipation" as both an historic event and a social process' are still experienced acutely (Beckles 2001: 90). The discourse on the past, and the present which has evolved from it, is infused with emotion. In fact, the past is so present that 'references to slavery and the ideologies/legacies of slavery are constantly used to frame and contextualise public debates and discussions' (Sheperd 2000: 57).

Though slavery was abolished over a century and a half ago, its legacy is felt clearly in Jamaica. Despite its national motto 'out of many, one people', Marcus Garvey's (1973: 5) observation in the early twentieth century – 'the [white, coloured and black] people mix in business but they do not mix in true society' – remains acute. Most Caribbean territories achieved statutory independence or at least large measures of autonomy in the second half of the twentieth century. Political independence having been achieved, the rejection of neocolonialism and the search for a national and cultural identity, which is so often linked to issues of 'race' and ethnicity (cf. Baronov and Yelvington 2003), continues particularly in the arts – music and literature as well as the fine arts (cf. Nettleford 1978; Ramchand 2000; Rohlehr 2000; Wheeler 1996). So the fact that a piece of art becomes the focus of so much debate is obvious rather than surprising. In addition, this artwork is in a public place and not in a museum or art gallery. This and the fact that its 'audience' is the general public, while it was created without previous public consultation, makes the sculpture subject to even more scrutiny, increasing the public scope of the debate.

---

[18] In fact, a previous sculpture contending for the Emancipation Park spot by A.D. Scott was met with so much disapproval that it was removed. The Bob Marley statue currently standing outside the National Stadium also had a predecessor which was rejected, purportedly because Marley's features were deemed 'too African'.

[19] 'Meaning of the statue', *Jamaica Gleaner*, 8 August 2003.

Commemorating slavery worldwide

The debate that burst forth in Kingston, though rooted in the local historical and social context, can be seen in the light of international efforts to come to terms with a past marked by slavery, slave trade and institutionalized racism.

Although a growing sensibility towards slavery and exploitative race relations has emerged in the United States (Drescher 2001),[20] Europe is definitely lagging behind in terms of responsiveness towards these issues, in public discourse as well as in physical commemoration. This is exemplified by the case of the Dutch slavery monument. This sculpture was erected in a park in Amsterdam in 2002, and the opening ceremony was attended by the Queen and national and international officials. However, the general public, consisting mainly of Afro-Caribbean Dutch, was refused entrance – policemen and tall gates prevented them from catching even a glimpse of the ceremony and they were forced to watch the proceedings on a large screen in an adjacent area. Needless to say, this did not go over too well with those for whom, ostensibly, the monument was created.

Caribbean societies have varying approaches towards historicizing slavery and emancipation, sometimes endorsing silence or forgetting rather than remembrance (Césaire 2001; Fiet 2001), sometimes utilizing the past to strengthen the process of nation-building (Beckles 2001). In the case of Redemption Song, the Jamaican government attempted to create a symbol strengthening national unity. Instead, a national dispute ensued, revealing fault lines of colour and class.

Emancipation, whether viewed as an event or a process, functions as the borderlands between the past (slavery and its successors: colonialism and oppression) and the future: physical freedom, political self-determination and cultural or 'racial' self-confidence. As with all borderlands, the construction of this symbolic domain remains contested. To many, it is tempting to continue looking back, depending on the 'legacy of slavery' to explain social ills, and at times succumbing to the seductive position of victimhood. While justifications for this course of thought are apparent, possibilities of agency and change are limited by focusing exclusively on history and the undeniably appalling injustices done in the past. Given this, it was a courageous move for both the Jamaican government and the sculptor to undertake the creation, or production, of a national symbol that, though

---

[20] Resulting largely from the civil rights movement and witnessed for instance by the annual Black History Month, affirmative action policies and the debates on politically correct language and on remuneration to the descendants of slaves.

flawed, focuses attention and hope on emancipation as a process, or a project if you will (cf. Sheperd 2000), that attempts to make sense of the past by looking forward. Unfortunately, the politics of race are played out at different levels. Laudable as this didactic effort was, it did not reckon with sensitivities of a population that is preoccupied with looking back.

References

Austin-Broos, D.J. (1997) *Jamaica Genesis: Religion and the Politics of Moral Orders.* Chicago: University of Chicago Press.

Barnes, N.B. (1997) 'Face of the nation: Race, nationalisms and identities in Jamaican beauty pageants.' In: C.L. Springfield (ed.), *Daughters of Caliban: Caribbean Women in the Twentieth Century*, pp. 285–306. Bloomington: Indiana University Press.

Baronov, D. and K.A. Yelvington (2003) 'Ethnicity, race, class and nationality.' In: R.S. Hillman and T.J. D'Agostino (eds), *Understanding the Contemporary Caribbean*, pp. 209–38. Kingston: Ian Randle.

Beckles, H. (2001) 'Emancipation in the British Caribbean'. In: G. Oostindie (ed.), *Facing up to the Past: Perspectives on the Commemoration of Slavery from Africa, the Americas and Europe*, pp. 90–4. Kingston: Ian Randle.

Bryan, P. (2000) *Inside Out and Outside.* In: *Factors in the Creation of Contemporary Jamaica.* Kingston: Grace Kennedy Foundation.

Césaire, I. (2001) 'To each his commemoration.' In: G. Oostindie (ed.), *Facing up to the Past: Perspectives on the Commemoration of Slavery from Africa, the Americas and Europe*, pp. 56–7. Kingston: Ian Randle.

Charles, C.A.D. (2003) 'Skin bleaching, self-hate and black identity in Jamaica.' *Journal of Black Studies*, vol. 33, no. 6, pp. 711–28.

Clarke, C.G. (1975) *Kingston, Jamaica: Urban Development and Social Change 1692–1962.* Berkeley: University of California Press.

—— (1985) 'A Caribbean creole capital: Kingston, Jamaica (1692–1938).' In: R.J. Ross and G.J. Telkamp (eds), *Colonial Cities: Essays on Urbanism in a Colonial Context*, pp. 153–70. Dordrecht: Martinus Nijhoff.

Cooke, Melville. (2003) 'Redemption Song vs. my Ms Jamaica.' *Jamaica Gleaner.*

Cooper, C. and A. Donnell (2004) 'Jamaican popular culture.' *Interventions: International Journal of Postcolonial Studies*, vol. 6, no. 1, pp. 1–17.

Drescher, S. (2001) 'Commemorating slavery and abolition in the United States of America.' In: G. Oostindie (ed.), *Facing up to the Past: Perspectives on the Commemoration of Slavery from Africa, the Americas and Europe*, pp. 109–12. Kingston: Ian Randle.

Fiet, L. (2001) 'Puerto Rico, slavery, race: Faded memories, erased histories.' In: G. Oostindie (ed.), *Facing up to the Past: Perspectives on the Commemoration of Slavery from Africa, the Americas and Europe*, pp. 70–4. Kingston: Ian Randle.

Garvey, M. (1973) 'The race question in Jamaica.' In: D. Lowenthal and L. Comitas (eds), *Consequences of Class and Color: West Indian Perspectives*, pp. 4–12. New York: Anchor. [Original edition: 1916.]

Hope, D. (2004) 'The British link-up crew.' *Interventions: International Journal of Postcolonial Studies*, vol. 6, no. 1, pp. 101–17.

Jamaica Gleaner

Jamaica Observer

Johnson, H. (1998) 'Introduction.' In: H. Johnson and K. Watson (eds), *The White Minority in the Caribbean*, pp. ix–xvi. Kingston: Ian Randle.

Moser, C. and J. Holland (1997) *Urban Poverty and Violence in Jamaica*. Washington DC: World Bank.

Nettleford, R. (1973) 'National identity and attitudes to race in Jamaica.' In: D. Lowenthal and L. Comitas (eds), *Consequences of Class and Color: West Indian Perspectives*, pp. 35–55. New York: Anchor.

—— (1978) *Caribbean Cultural Identity: The Case of Jamaica*. Los Angeles: CAAS.

NRCA (National Resource Conservation Agency) (1998) *Jamaica State of the Environment Report*. Kingston: NRCA.

Patterson, P.J. (2002) Address By Prime Minister Rt. Hon. P.J. Patterson, QC, MP at Opening of Emancipation Park, July 31, 2002. [http://www.pnpjamaica.com/speeches oct5a.htm.]

Pruitt, D. and S. LaFont (1995) 'For love and money: Romance tourism in Jamaica.' *Annals of Tourism Research*, vol. 22, no. 2, pp. 422–40.

Ramchand, K. (2000) 'The lost literature of the West Indies.' In: K. Hall and D. Benn (eds), *Contending with Destiny: The Caribbean in the 21st Century*, pp. 514–29. Kingston: Ian Randle.

Randall, S.J. (2003) 'The historical context.' In: R.S. Hillman and T.J. D'Agostino (eds), *Understanding the Contemporary Caribbean*, pp. 51–83. Kingston: Ian Randle.

Rohler, G. (2000) 'Change and prophecy in the Trinidad and Tobago calypso towards the 21st Century.' In: K. Hall and D. Benn (eds), *Contending with Destiny: The Caribbean in the 21st Century*, pp. 542–74. Kingston: Ian Randle.

Sheperd, V.A. (2000) 'Image, representation and the project of emancipation in the Commonwealth Caribbean.' In: K. Hall and D. Benn (eds), *Contending with Destiny: The Caribbean in the 21st Century*, pp. 53–79. Kingston: Ian Randle.

Stewart, K. (2002) '"So wha, mi nuh fi live to?": Interpreting violence in Jamaica through the dancehall culture.' *Ideaz*, vol. 1, no. 1, pp. 17–28.

Stolzoff, N. (2000) *Wake the Town and Tell the People: Dancehall Culture in Jamaica*. Durham: Duke University Press.

Thomas, D.A. (2002) 'Modern blackness: "What we are and what we hope to be".' *Small Axe*, vol. 12, pp. 25–48.

Wheeler, E.A. (1996) *Unthinkable Cities: Kingston and Los Angeles*. University of California, Berkeley: Ann Arbor, University Microfilms International. [PhD thesis.]

Chapter 5

POWER OF REMEMBRANCE, SUPREMACY OF OBLIVION
History of the 'National Monuments' in Ljubljana

Božidar Jezernik

Introduction

According to Robert Musil, the most obvious comment to make about public monuments is that people do not notice them: though erected to attract attention, they are in fact attention-proof (cit. Hojda and Pokorný 1996: 9). Attention focuses on them only at the critical moments when they are being erected or removed. Otherwise, people appreciate them in a state of 'distraction', as they pass by, unlike the 'concentrated' absorption given to statues in a gallery. Response to monuments is obviously not intellectual or positivist; rather, it involves fantasy, wish processes and dreams (Urry 1996: 51). Like all symbols, statues and the ideologies they represent organize people's experience and express relations between groups. The narratives people attach to statues give meaning to their identity. Contrived to 'speak' to the proverbial man in the street, they serve as a school of patriotism, open to everybody. 'Until recently,' suggests David Lowenthal, 'most monuments were exhortations to imitate the virtues they commemorated; they reminded people what to believe and how to behave' (Lowenthal 1985: 322).

Public monuments are, however, intended not so much to make the public learn something but to make them *become* something: members of a community. When, in the second half of the nineteenth century, Ljubljana's German façade began to be replaced with a Slovenian one, the local bourgeoisie attempted to arouse mass allegiance to icons of new (Slovenian) collective identity. In the ensuing atmosphere, public monuments proliferated: those thereby recognized as 'great men' and the ideas which they stood for were appropriated as the spiritual heritage of the Slovenian nation and were instrumental in creating the notion of community while suppressing local memories. One of the main proponents of monument construction, Lovro Toman, wrote in 1850: 'Every monument erected by the nation to men of excellence is erected to the nation itself, because it is a sign that the nation has become aware of its duty, its calling, its power, that it has recognized the significance of such a construction as a token of gratitude and as an encouragement to similar acts of merit' (Kamnogorski 1850: 61).

The past embodied in the statues of eminent citizens was needed to strengthen the purpose of those who strived to awaken the national consciousness to evaluate Slovenian culture by establishing what was notable. Through 'national monuments', the objectification of national spirit and the recognition of the people of a certain nation as embodiments of that spirit took an enduring form that emphasized long-term continuity. They symbolize and memorialize the continuity of communal ties: as such, they were instrumental in the process of cultural homogenization of the Slovenians and in the construction of boundaries between them and others. Essential to this process was the selection of 'heroes', of the time and place where the statues were erected, and of the artists who created them. They were meant to serve as a material expression and foundation of the Slovenian nation's self-confidence and as a testament to its importance and excellence. Thus, in 1865 the Maribor committee for the erection of the Bishop Slomšek statue addressed readers of the newspaper *Novice* as follows:

> All educated nations of the present and past have erected valuable public monuments to men who were their pride; the Slovenian nation must not lag behind in this admirable habit ... Let us show the world that we are worthy to live as a nation, that we feel like a nation and can, as a nation, join the circle of equal nations.

(Mariborski odbor 1865)

The so-called national monuments served as a palpable connection with people's historically significant figures. They served the purpose of organizing people's relationship with the past, giving people a sense of history. Without them, Slovenians would be like 'a nation without history', because they would supposedly lack both a sense of its own identity and an awareness of its place in the temporal scheme of things. The proponents of construction of statues dedicated to eminent Slovenians used the past in the service of their future aims and as its guide. They presented new ideas and values and filled them with emotion; they produced new interpretation of the world which, eventually, became the basis of perception and action. By means of 'national monuments', certain individuals were made the expression of the nation that could be gazed at, learned from and, eventually, fought for. As Davorin Trstenjak wrote:

> We have to demonstrate that, like other nations, we, too, respect the men who with their radiant spirit and mighty words penetrated so deep into the soul of their nation that it became aware of and began gloriously to cultivate its spiritual powers. Remember the popular saying: He who does not respect himself gains no respect from others.

(Trstenjak 1858: 1–2)

A town to shame the whole world

As a rule, resurgent national and ethnic alliances require symbolic links with the past which serves as a living spring supplying the essential elements of national mythologies (see Lowenthal 1985: 396, 1996: 58). During the nineteenth century, Slovenian nationalism was emphatically apolitical and concentrated its efforts on the cultural field, giving priority to men of letters such as the first Slovenian poet, Valentin Vodnik (1758–1819), foremost Slovenian poet, France Prešeren (1800–1848), and the creator of the Slovenian written language, Primož Trubar (1508–1585). The famous American war correspondent, William Shirer, who visited Ljubljana on 10 March 1938, called it 'a town to shame the whole world'. He said that Ljubljana was 'full of statues and not one of them of a soldier. Only poets and thinkers have been so honoured' (Shirer 1988: 95). Shirer cannot have realized that the Slovenian liberal bourgeoisie were using culture as part of a strategy to assure people who had no political definition before 1848 that they did, indeed, enjoy a national identity. Likewise, it seems, he jumped to conclusions about Ljubljana's statues before he had seen all, for up to the end of the 1880s the most popular statue in the city was that of the Austrian field marshal, Count Johann Joseph Radetzky (1766–1858). Marshal Radetzky had close links with Slovenia: he was married to Countess Frančiška Strasold, a native of Carniola, and inherited considerable property there; he thus acquired Carniolan citizenship in 1807. He was very popular with Slovenian soldiers who placed full reliance upon him. His exploits were celebrated in folk songs such as, for example:

Radetzky is a real gentleman,
He will prevail every time, everywhere ...

(Mal 1928: 714)

In 1852, the people of Ljubljana erected a life-size statue of Count Radetzky in marshal's uniform in Zvezda Park. As the inscription on the pedestal explained, the statue portrayed him during the battle against the Italians of 1849, encouraging the soldiers with the promise that the battle would be short. Six years later, the municipality decided that this statue was not really a worthy monument, because it was only a cast. They therefore commissioned the famous Viennese sculptor Anton Dominik Fernkorn to produce a bust of Radetzky. The full-length statue was stored in the municipal depot where it remained until it was re-erected in front of Tivoli Castle in 1882. The bust by Fernkorn combined strong realism in its portrayal of the subject – his uniform and medals – with romantic fervour – he was depicted crowned with a laurel wreath symbolizing victory and triumph (Figure 1). The costs of restoring the monument were not insignificant but with it Ljubljana regained its first representative public monument and its citizens were fond of it (Čopič, Prelovšek and Žitko 1991: 24–5).

The ideals of official memory must be enacted in ceremonies and commemorations if they are to last (Shackel 2001: 660). Accordingly, Radetzky's monument was made a focal point for many events and acts that extolled the capital of the Duchy of Carniola and symbolized its loyalty to the Habsburg crown. On the other hand, this ceremonial role also made the monument a venue for various merry bands of students who played many night-time pranks there. In one of these, they piled all the benches in Zvezda Park one on top of the other around the Radetzky statue, making a ladder onto it. Two of their number climbed up and sat astride each of the old Austrian field marshal's shoulders and slapped his face from either side. The rest of the company made an appreciative audience. Of course they all ran away when they heard the police coming (Tuma 1937: 62).

Public memory is more a reflection of present political and social relations than a true reconstruction of the past. As conditions change socially, politically and ideologically, the collective memory of the past also changes (Teski and Climo 1995: 2).

Figure 1 Count Radetzky (Source: Historical Archive of Ljubljana).

That is why, after a revolutionary coup, certain monuments no longer fit the new historical and ideological context of a society. The demolition of such monuments is usually an aspect or even a trigger of historic change. When democratic systems were introduced into most European countries the old symbols of monarchy were discarded (see e.g. Firth 1973: 328–67). The most spectacular example of the demolition of old monuments and the erection of new ones took place when Bastille was stormed in 1789, marking the beginning of the French Revolution. Correspondingly, Radetzky's destiny as a Slovenian hero was short-lived. During the night of 30 December 1918, two 'patriots' pulled down the statue in front of Tivoli Castle and the same night his bust was removed from Zvezda Park.

Themselves voiceless, public monuments required interpretation. Radetzky's statues in Zvezda Park and in front of Tivoli Castle were expressions of the Austrian patriotism of the Slovenian and German population of Ljubljana. But the situation changed with the next monument, which was dedicated to the Carniolan Anastasius Grün, Anton Alexander Count Auersperg (1806–1876), a poet and politician whose initial support for Slovenians gave way in his old age to support for the predominance of the German language and culture in all Austrian lands. Slovenians would gather around the monument to Valentin Vodnik (Figure 2) and Germans around the monument to Anastasius Grün. In 1862, when Germans were still in the majority in the municipal council, Grün was declared an honorary citizen of Ljubljana, but by 1886 circumstances had changed considerably and the municipal council refused to participate in the celebration of the anniversary of his death. The councillors expressed their respect for Anastasius Grün, the poet, but stated that there were no cogent arguments for their involvement in the celebration of his political activities (Čopič, Prelovšek and Žitko 1991: 28; Kos 1997: 14).

As we always define ourselves in relation to others, our identity depends on these others and their identity. When important issues and interests are involved in a social situation, the division between Us and Them may lead to antagonism and conflict, and action is likely to follow the mental process of identification (Mach 1993: 10). This process is facilitated by the fact that in asserting our virtues, we harp on others' vices. In fact, on various occasions, Germans decorated the monument to Grün with flowers and garlands tied with ribbons in the German national colours; however, Slovenians saw such acts as provocations to their national feelings and sometimes responded by besmearing the monument with dirt. Several such events were reported in Ljubljana newspapers (see e.g. *Slovenec*, 'The Slovenian', 14 March 1904, 11 April 1906 and 17 September 1908; *Slovenski Narod*, 'The Slovenian Nation', 11 April 1906). Soon after the First World War and collapse of the Dual Monarchy, the portrait medallion was removed and in 1924, on the occasion of the first Marianic Congress which was held in Ljubljana, the Teutonic Knights filled the empty space with the bas-relief of the Virgin and the Child which is still there (Čopič, Prelovšek and Žitko 1991: 28).

Figure 2 Valentin Vodnik (Source: Historical Archive of Ljubljana).

'Spiritual monuments' versus 'Stone or iron manikins'

While a succession of 'national monuments' constantly changed the appearance of nineteenth- and twentieth-century Ljubljana, their principal function was to forge the new collective identity supposedly handed down by history. The main stages in this identity were, first, modernization led by the Slovenian liberal bourgeoisie and, second, half a century later, communist revolution. Both were radical breaks with history. Erecting monuments to Vodnik, Prešeren and Trubar (Figure 3) was part of Mayor Ivan Hribar's efforts to modernize the city, ensure its economic and cultural development and improve the standard of living. By playing the culture card, however, Hribar and his Liberal municipal council hoped to strengthen Ljubljana's Slovenian identity and its role as the capital of the Slovenian national territory. Upon entering office in 1896, Hribar stated that as the supreme administrator of the city it was his 'sacred duty' to take into account the fact that the vast majority of Ljubljana's population was Slovenian. It is well known that identification with a national past often serves as an assurance of worth for the subjugated or bolsters a new sovereignty: people deprived by conquest of their proper past strive to retrieve the reassurance of its memory (Lowenthal 1985: 44). It came as no surprise when, in 1903, his committee for the erection of the Prešeren monument appealed to the public by asking them: 'Will not this monument to Prešeren constitute public proof, concrete proof, that our white Ljubljana is Slovenian?' (Kos 1997: 19–20).

However, nationality is a political phenomenon, especially when it is used in a struggle for power with the others who are identified as culturally alien (Mach 1993: 14). If the proponents of national 'awakening' were to construct a Slovenian nation, they had to stand for something, culturally and politically: this inevitably meant distinguishing Slovenians from the non-Slovenians. Accordingly, 'national monuments' did not simply put history on display; they were instrumental in creating a particular kind of history for the public and were lending that history their own legitimizing stamp. They gave the outward appearance of a new image of Slovenian community, glorious and enduring, with a historical mission and rights in relation to other groups. The men of letters were valuable not only as symbolic proof of Slovenia's place in the family of civilized nations, but also as symbols of transcendence, able to displace attention from political conflicts within the group and project them on those outside it. The developing Slovenian bourgeoisie, which was not strong enough to compete with its German rivals, endeavoured to contain the German *Drang nach Ost*. With the aim of legitimizing their own position and justifying their own claims, they tried to take control over the past (cf. Ferro 1984: vii).

The first proposal to set up a memorial to Valentin Vodnik was made at the time of the celebrations for the centenary of the poet's birth. Matija Majar expressed his pleasure that a total of 1,100 gold dinars had been collected for the Vodnik monument, stating that it was 'a very auspicious sign that patriotism is not just a meaningless word in our parts', because the monument was to stand 'in honour and memory of the poet and for the benefit and education of the Slovenian people'. He maintained that the most suitable monument for this purpose would be spiritual in nature and not a structure of iron and stone:

> Of what benefit would it be if a statue was erected, even if it was to reach up to the clouds? It would be a dead monument. The merit of Valentin Vodnik is spiritual and therefore his monument should also be spiritual to strengthen the patriotism and education of our nation. The monument must not be dead, it must be alive, it must grow, blossom and bear fruit year by year, it must speak and remind us of Valentin Vodnik year by year, it must encourage our students to learn and love their nation and their Slovenian language.

(Majar 1858)

This could be achieved, according to Majar, if the annual interest from the sum collected was spent on books to be awarded at a special ceremony to the best students after a speech glorifying Valentin Vodnik. By this means, the Vodnik monument would be 'alive and spiritual,' whereas 'stone or iron manikins' would only benefit the stonecutter or blacksmith (Majar 1858).

Some other participants in the debate shared this point of view. K. Žavčanin, for instance, maintained that anyone interested in the work of great Slovenian men 'will not go around enquiring about these men in squares and streets; he will go to the counter of our book vendors, he will go to our antiquarians, he will go to our archives, he will go to our museums!' (Žavčanin 1858). But for those who thought that the monuments are powerful because seeing is believing, a 'spiritual monument' was clearly not enough. Thus, Josip Novak (1858), for instance, rejected Majar's opinion on the basis that 'Other nations have also erected such monuments to their great men and nobody has ever accused them of extravagance. A magic spirit surrounds this kind of monument, encouraging noble and gentle thoughts in the educated man who stands in front of an image of a famous individual.'

Fund raising for the monument to Valentin Vodnik began in 1858, but despite great patriotic enthusiasm, the sum was collected only in 1889. It was raised mainly at special events where national feeling was expressed by singing national anthems and flying the national flag, reciting poetry, creating *tableaux vivants* and in many other ways. The organizing committee for the erection of the Vodnik monument decided that the first Slovenian national monument in Ljubljana must be the product of a Slovenian mind and Slovenian hands in order to assert the value of Slovenian creativity. The

Figure 3 Primož Trubar (Source: Historical Archive of Ljubljana).

area in front of the entrance to the large secondary school that stood in present-day Vodnik Square was selected as the site of the monument to emphasize his role in the development of Slovenian schools. It was envisaged that 'Slovenian youth shall always gaze upon the one who inspired the nation for a great cause' (Levec 1888: 765 and 1889: 446).

Blut und Boden

Slovenian 'national monuments' were thus works of art in their own right but also 'more than' art. Their task was to incarnate the Slovenian nation, its perseverance and steadfastness, and, conversely, to serve as visible guarantors of national identity. Lovro Toman proposed the first monument to France Prešeren in 1850 (Figure 4), only a year after the poet's death. Eventually, the campaign for erecting it gained support from most strata of the Slovenian public. In Vienna in 1890, Matija Murko called on Slovenians not to spend their energy on monuments to locally important men, but rather focus their attention on those who were 'the true pride of all Slovenians'. He maintained that the Prešeren monument represented 'a new sign of the spiritual unity of our nation, which has been divided by history and circumstances'. Since one of the main roles of the 'national monuments' was to divide the social world into good and evil by defining Slovenia's enemies and directing the nation's thought, emotions and actions against them, he associated it with the idea that Slovenians were 'a robust bulwark against Germans who covet the Adriatic Sea' (Murko 1891: 81–7).

Such ideas were quite widespread in nineteenth-century Europe. The British sculptor Benjamin Robert Haydon, for instance, wrote in his diary in 1808 that he had made a sketch of an idea that struck him while at Dover, 'of a colossal statue of Britannia and her Lion, on Shakespeare's Cliff, right opposite the coast of France' (Taylor 1853: I, 108).

When a national community has its own land, it constitutes the centre of the world and is endowed with a sacred quality. A national community does not only claim its right to inhabit a given territory, but also demands that this right should be exclusive or, at least, that it alone should have sovereignty over the land (Mach 1993: 173–4). Fund raising for the Prešeren monument thus soon turned into a political campaign or, more precisely, became a political campaign employing cultural means. This was clearly apparent when the ceremony of the unveiling of the monument took place on 10 September 1905. The newspaper *Slovenski Narod* published an article addressed 'To Esteemed Guests', which was probably written by Ivan Hribar. The author said that by coming to Ljubljana, the Slavic guests (the mayors of Slavic capitals and representatives of Slavic universities, academies and various societies) showed that the Slovenian nation was not alone in its struggle and that all Slavs were like a 'large family bound into an indivisible whole by the bonds of fraternity and love'. He also expressed the hope that, strengthened by the 'moral and material' support of their Slavic brothers, Slovenians would become masters of their land.

Here, next to the monument of a spiritual giant who long ago became a son of all Slavic nations, we are bound by even greater bonds that bring the small Slovenian nation closer to its greater and more fortunate brothers; may the mutual love and noble idea of Slavic unity grow into a mighty flame, from which the Slovenian nation will draw new strength for the heavy battle fought with numerous adversaries.

In their addresses, Slavic speakers emphasized Slavic unity, expressing their support for Hribar's political endeavours. After a procession of participants, a crowd gathered around the veiled monument. After a performance by a choir, the special speaker Dr Ivan Tavčar addressed those present. He compared Prešeren to a giant oak whose roots are embedded deep in the soil and whose crown reaches the sky. He stated that the unveiling of the monument 'is also a celebration of pan-Slavism, a fact we do not hide and would not wish to hide from anyone!' He declared that the monument was erected by the Slovenian nation and was bequeathed to Ljubljana by Slovenian hearts.

And thus the monument to France Prešeren will stand in the heart of our lands! It will be a guardian of the Slovenian essence of these lands! It must not be spoilt by the passage of time or by rust! We place it in the hands of our progeny as a sacred heritage and they must defend it like an army that repels all assaults on their fatherland both from the north and from the south!

(Kos 1997: 135)

Until the end of time

It was mainly political ambitions that made Mayor Hribar join those in favour of the erection of the monument to the Austrian emperor Franz Joseph I (1830 – 1916) (Figure 5). In 1895, the year of the great earthquake, when the emperor visited the devastated city, the city authorities of Ljubljana decided to erect a monument in the capital of the province as a visible and permanent expression of gratitude for the emperor's kindness in visiting them (Hribar 1903: 1). According to Hribar, the main reason for the erection of the monument to Franz Joseph was to 'make the emperor interested in the Slovenian nation and turn him into a patron of Slovenian national idea'. Hribar also pointed out how the idea paid dividends:

Figure 4 France Prešeren (Source: Historical Archive of Ljubljana).

When the monument was erected, Emperor Franz Joseph, who had never been particularly generous to the wishes of our nation, donated 10,000 crowns towards the construction of the German theatre, that strategic stronghold of the German spirit in the country, which was originally artificially kept alive by the Savings Bank of Carniola. The sum was not great. Actually, it was a pauper's present from a rich emperor; but it sufficed to show clearly the value of his constantly repeated phrase that he loved all 'his' nations equally (Hribar 1984: II, 114–15).

The Ljubljana municipal council allocated the sum of 20,000 crowns to the construction of the monument. At an assembly of Slovenian mayors on 18 August 1898 it was decided that all Slovenian municipalities in Carniola, Styria, Carinthia and the Littoral should contribute to the costs of this monument, 'so that the Slovenian nation could express its feelings of gratitude to that ruler, under whose reign the nation began a new life' (Hribar 1903: 1). In 1903, a competition was launched which only Slovenian and Croatian sculptors could enter and which was won by Svetoslav Peruzzi. Peruzzi's design for the monument was very simple: a bust of the emperor and a kneeling female figure representing Ljubljana were placed on a prismatic pedestal. Many contemporary experts maintained that the city authorities, that is their committee, opted simply for the cheapest offer which was, as *Slovenec* (The Slovenian), the daily with the largest circulation, put it on 7 January 1904, 'not monumental at all'. On 8 March 1904 the same newspaper complained bitterly:

> We are a small nation with an unfortunate fate. Adversaries will not recognize our literature, and now we are about to put a weapon in their hands ourselves, so that they can also reject our art, by refusing a good work that would be our pride – and by erecting statues that will speak of our poverty to the future generations. It would be better if we engraved all our monuments with the following words: It is miserable, because such a meagre sum was allotted for its making! Art is no exception: *aut Caesar aut nihil*!

The unveiling of the emperor's monument was lavishly celebrated in the presence of the representatives of ecclesiastic, secular and military authorities, the Ljubljana municipal council, the fire brigade, the philharmonic and the wardens of the judicial administration. A military brass band played. The event was also attended by the Prince Bishop Jeglič, who was 'repelled' by the thought of finding himself in the company of 'supporters of the Liberal Party', but he could not risk 'appearing unpatriotic' (Jeglič 1908: 47).

Figure 5 Franz Joseph (Source: Historical Archive of Ljubljana).

The following year, the emperor's monument in front of the courthouse featured significantly in the election campaign. On 17 April 1911 Dr Šuštaršič addressed a rally of the Slovenian People's Party in the Union Hall. According to a report summing up his speech in *Slovenec* the following day, monuments were the clearest indication of the impact of the Liberal Party in Ljubljana:

Just look at the emperor's monument! If I were the government, I would prohibit such a monument to our emperor. (Hear, hear! It's a scandal!) This monument shames the emperor! (Hear, hear! It is true!) Is that the emperor? I once knew an old porter who looked like that 'emperor', placed by the Liberals in front of the courthouse. (Laughter.) If our grandchildren see that monument, they will get a completely false idea of what our emperor looked like. One of the first proposals to the future municipal council is to remove the 'emperor's' statue that now stands in front of the courthouse, and place it in the museum. In Hribar's corner in the museum! (Loud laughter.) A sign will have to be placed below the statue that will say: This 'emperor' was made in Ljubljana when Ivan Hribar was the mayor. (Very loud laughter.)

In turbulent times the monument itself was frequently a target for those who thought that they had something important to say and that a monument to the emperor was the most appropriate object to convey their message. In 1910, for instance, a 'special vase', or in plain language a chamber pot, was discovered at the spot on the Emperor's birthday (18 August) (Hribar 1984: II, 115).

Another monument to Emperor Franz Joseph I was erected at Ljubljana Castle during the First World War to mark the emperor's eighty-sixth birthday. In the circumstances, the monument instilled even profounder patriotism. On 18 August 1916 *Slovenec* issued a long report on the celebration in Ljubljana and the unveiling of the emperor's monument at Ljubljana Castle, which took place 'without special pomp but with profound inner emotion and great pride. Yes, pride in the fact that we are Austrians, sons of a country that lovingly cares for all its nations like no other'. The monument was embraced with iron chains as a symbol of 'our will of steel in this age of steel'. The mayor of Ljubljana, Dr Ivan Tavčar, placed the monument in the municipal care and stated in Slovenian and German:

On behalf of the provincial and diocesan capital of Ljubljana I declare that our municipality hereby takes this monument, erected on its land, into its care and I pledge that the city will until the end of time lovingly and conscientiously guard this symbol of dynastic loyalty of the population of this duchy (Cesarjev 1916).

As argued by Shackel (2001: 657), public memory associated with highly visible objects is constantly reconstructed, changed and challenged. Accordingly, 'the 'indescribable joy' which filled Slovenian hearts during the First World War dried up as soon as hostilities were over: anonymous patriots tore down both imperial monuments.

Equestrian monuments

The First World War left masses of dead and wounded. The bloodbath ended in major changes to the geopolitical map of Europe, one of the most visible of which was the collapse of the Dual Monarchy. This put an end to any gratitude towards the emperor: it was decided that the monument should be pulled down and replaced, in 1926, with a monument to Fran Miklošič (1813–1891), who was regarded as the greatest grammarian of the nineteenth century. This famous scholar of Slavic languages was particularly popular among Slovenians after the war because during his lifetime he had worked for the union of all Slovenians and for linguistic equality in schools and offices. But the elders of Ljubljana, for reasons of economy, decided simply to set the bust of Miklošič on the pedestal which had originally supported the memorial to the emperor, whose name was obliterated. The stonecutter chiselled a new inscription. What remained on the pedestal of the Miklošič memorial were two imperial eagles and the kneeling woman holding the coat of arms of the city of Ljubljana in one hand and in the other a laurel wreath which she raises towards the figure above her. On the back of the memorial is a bas-relief of victims of the earthquake in 1895 (Čopič, Prelovšek and Žitko 1991: 84).

After the disintegration of Austria-Hungary, most Slovenian territory was incorporated into a new state, the Kingdom of Serbs, Croats and Slovenians, which united the majority of the southern Slavs. In a country shared with their Slav compatriots, Slovenians found it easier to breathe, for after many centuries of domination they were free from Austria-Hungary, now referred to as the 'prison house of nations'. They concluded with satisfaction that the new state had not only liberated them from a 'centuries-long yoke' but had also made their ancient dreams come true. The deep emotions of the people around this time found their expression in a monument to a man who symbolized the realization of their dreams, King Peter I (1844–1921), called the Liberator. Equestrian monuments were considered as the most appropriate mark of respect that can be paid by a people to their sovereign (see e.g. Cleghorn 1824: 170).

The monument was to last for 'all eternity' and was therefore cast in bronze and erected in front of the town hall (Figure 6).

After the death of King Peter, the throne went to his son, Alexander I (1888 – 1934), who transformed the Kingdom of Serbs, Croats and Slovenians into the Kingdom of Yugoslavia, so earning the title 'the

Figure 6 King Peter (Source: Historical Archive of Ljubljana).

Unifier'. As stated in 1940 by Josip Pipenbacher, president of the Committee for the Erection of the Monument to the Noble King, Alexander the Unifier had 'decided to build what had been the dream and wish of all noble Yugoslav men for centuries, that is the national unification of all Yugoslavs' (Maleš 1940: 7). He, too, deserved a monument and one was erected in his honour in Ljubljana. The execution of the monument was entrusted to the most distinguished Slovenian sculptor, Lojze Dolinar, who was well prepared for the public tender for the erection of the monument in Ljubljana, having already made detailed studies of an equestrian figure when he received the commission for the earlier monument to King Peter. He studied the image of the horse for many years, attempting to create a perfect harmony between its shape and that of the rider, and producing an accurately sculpted form. Although made in style of already obsolete European models, the sculpture was admired by all. Its height, six metres on top of a four-metre pedestal, made it 'the greatest monument in Slovenian territory' in its dimensions and its artistic merits (Maleš 1940: 19). The new equestrian monument aroused much enthusiasm. Thus Martin Bečina, for instance, declared:

> There is no doubt that with Dolinar's monument to Alexander I, Slovenians have received a work of art of such high quality that it will remain a testament to the creative powers of the Slovenian nation for many future generations and will evoke the live connection between our artist and the all-pervading life pulse of the nation to historians who will in the distant future trace our authentic national tenor in surviving artistic documents.

> (Maleš 1940: 20)

The memory of certain historical personages and their deeds is, on the one hand, a witness to the community and its continuity in space and time, that is the nation, and on the other hand, it is a witness to territorial validity and scope of a certain political influence. King Alexander on a mount symbolized the stability and security of the national borders, which led some to think of erecting a similar statue in Maribor to guard the northern border. Alexander I was assassinated in Marseilles on 9 October 1934. An announcement by the Committee for the Erection of the Monument, addressed to all Slovenians and published on 1 December 1934, claimed:

> To this embodiment of our will, to this glorification of our name and to this tried and tested faith, we wish to give a visible expression and a permanent record by erecting a worthy artistic monument to Him in Ljubljana and Maribor... His form, carved in marble and bronze, shall be a sign of victory over the power of Death. In Ljubljana, the third royal city of the state and the capital of Drava Province, which binds and connects all arteries of our cultural and economic life, it shall stand as a reflection and expression of freedom secured with His assistance, which unleashed the potential of all our activities, long suppressed by foreign forces. In Maribor, in that stronghold on our northern border, once stolen and retrieved by the joint forces of this nation, it shall be an eternal sentinel, in the shelter of which the freed nation will peacefully enjoy the

blessed fruit of its fatherland. And it shall also admonish: No entry!

(Maleš 1940: 21)

The equestrian monument to King Alexander I was unveiled on 6 September 1940 (Figure 7). In the year of the erection of the monument to the Noble King, Europe was already engaged in a new war. On 6 April 1941, without warning, the Axis Forces attacked the Kingdom of Yugoslavia from different directions and its army capitulated soon afterwards. Occupied Yugoslavia was divided among the aggressors. Slovenian territory itself was parcelled out to three occupying forces (Germany, Italy and Hungary), with the capital, Ljubljana, in the zone occupied by the Italians. Though astonished by the 'high level' of culture they encountered there, they found Ljubljana, to the discontent of its inhabitants, not big enough for the equestrian monuments. On 3 July 1941, High Commissioner Emilio Grazioli himself made an attempt to appease the representatives of Slovenian political circles who were present at the session of his council by explaining that the monuments to the 'Serbian' kings would be kept in a museum as 'works of art' (Grazioli 1941: 2). In fact, both monuments were pulled down by the Italian occupying authorities during the summer of 1941. They performed this part of their 'cultural mission' thoroughly: one night in early August, the pedestal of Alexander's monument was destroyed with a concrete crusher, and the stones and rubble were taken away in a lorry (Šnuderl 1993: 108, 132, 140). The feelings aroused in the inhabitants of Ljubljana by this act were noted by a contemporary in his diary:

Nobody knows if the matter will be brought up again. But by pulling it down, they have offended Slovenian national feelings, not dynastic ones. The statue was beautiful and served as a decoration. It was ours. Its removal is a blow to Slovenians. Why was it in the way? Would not it have been better for them if they had left it, with a fine smile on their lips, in generous awareness of their power and force? This way, they performed the act of a weakling and lost all sympathy.

(Šnuderl 1993: 132)

When the war was over, royal monuments were already history. Only in 1954, on the then Liberation Square, where Alexander's monument once stood and where 'the empty space literally cried out for a new memorial', was a huge anchor from the Slovenian coast placed by the communist authorities on a low stone base to celebrate Slovenia's access to the sea (Čopič, Prelovšek and Žitko 1991: 16).

Figure 7 King Alexander (Source: Historical Archive of Ljubljana).

The ephemeral nature of bronze and marble

The Second World War lasted longer than the First, but its outcome was similar: masses of dead and wounded and new changes to the geopolitical map of Europe. The first Yugoslavia, which had in the meantime been identified by its populace as a 'prison house of nations', was replaced by a new state which proposed the final solution to social and national problems. The colossal undertaking was, of course, impossible without new heroes. As suggested by David Gross, the end of the Second World War brought with it a discrediting of statist memory, as the excesses of Fascism and National Socialism made it clear that 'too much politicization of memory was not a laudable goal' (2000: 121–2). But George Orwell's syllogism ('He who controls the past controls the future; he who controls the present controls the past') held a too powerful sway on the new elite who strived to transform yesterday's villains into tomorrow's heroes. The result was an increase in memorial sculpture, with an extensive new programme of monuments to the national liberation struggle and the socialist revolution. Three of the giants of the period were offered up to eternity in bronze and marble in the city: Josip Broz Tito (1892–1980) (Figure 8), Boris Kidrič (1912–1953) (Figure 9) and Edvard Kardelj (1910–1979).

The objectification of the national spirit and the recognition of the people of a certain nation as embodiments of that spirit must take an enduring form that emphasizes the long term and continuity. Bronze and marble may be 'eternal' materials, but their use-by date was much sooner than had been envisaged. As argued by Lewis Mumford, such materials give a false sense of continuity and a deceptive assurance of life. 'But the fact is that the exterior form can only confirm an inner life: it is not a substitute. All living beliefs, all living desires and ideas must be perpetually renewed, from generation to generation: re-thought, re-considered, re-willed, re-built, if they are to endure' (Mumford 1938: 435). The monuments that were supposed to last until 'the end of human history' survived a mere half a century. This short period of time sufficed to prove Tito's Yugoslavia to be a state in which the national and social questions were not resolved. As we know, the disintegration of the second Yugoslavia followed and five new countries were formed on its territory. In newly created countries, monuments to former heroes receive special attention. Radical changes in political systems as a rule bring about changes in the symbolic system of a state which consist in the removal of symbols of the old order and the introduction of those which represent the new ideology and the new political leadership. Thus, one of the first official acts of the new Slovenian government was to remove photographs of Josip Broz from government offices; his statues were also promptly done away with. The full-length statue which the sculptor Boris Kalin had carved in white marble in 1957 stood in the entrance hall of Slovenian Parliament House until 1990 when it was taken to the City Museum. The famous full-length figure of Marshall Tito made by Antun Augustinčić in 1947, a cast of which stood in front of the then Museum of Revolution, was removed in 1990 to an 'unknown location' (Čopič, Prelovšek and Žitko 1991: 26, 154).

In 1995 procedures for the removal of 'monuments to the labour movement, national liberation struggle and the construction of socialism' were initiated by the city council. These would eventually lead to the elimination of monuments to Boris Kidrič and Edvard Kardelj, persons who were central to the socialist revolution, 'in the name of which thousands of Slovenians were murdered'. Those proposing the move pointed out that the removal of these two monuments would not mean their destruction but merely 'clearing of the public areas in question' and putting them 'into a place where history is preserved, a museum' (Kovačič 1996: 30). Or, according to a statement made at the meeting of the city council by a contemporary Slovenian writer and poet, Jože Snoj, 'the monuments to criminals and dictators must finally be removed from public areas and placed in a museum to serve for the studies of historians and art historians' (Žolnir 1996).

The proposal to remove the Kardelj and Kidrič monuments triggered an extremely lively debate. The main Slovenian daily newspaper *Delo* even started a special column, 'Monuments Yes, Monuments No' on the letters to the editor page. Many contributors agreed with the proposal, claiming that 'even in the former Soviet Union, the communist monuments to Stalin and other criminals have been removed' (Stibrič 1996: 16). But on the other hand, many passionately opposed 'such vandalism', especially former partisans, who strongly objected to the destruction of monuments to their wartime leaders. On behalf of Second World War Liberation Front veterans, Niko Lukež raised the following questions:

> Are Edvard Kardelj and Boris Kidrič so controversial because they were co-founders and organizers of the Liberation Front of the Slovenian Nation, which took part in the antifascist and anti-Nazi international democratic struggle as based on the Atlantic Charter, fighting for the survival of the Slovenian nation and for a united Slovenia(Liberation Front programme) through the formation of the Slovenian state, the decisions of the first Slovenian parliament, the assembly of the Slovenian delegates and deputies in Kočevje and the second conference of Avnoj? Is it highly controversial that Boris Kidrič was the first president of the Slovenian government in Ajdovščina, which at the time was still under Italian rule?

Conclusion

Paul Connerton (1989) argued convincingly that we remember not only individually but also collectively, as members of societies. Later, John Urry (1996) made it clear that people not only remember but also forget together. For Urry, forgetting is as socially structured a process as remembering. Social memories are often organized around particular artefacts, monuments,

Figure 8 Josip Broz Tito (Source: Historical Archive of Ljubljana).

etc. Thus, the erection of a monument makes a basis for remembering and its removal makes a basis for forgetting (Urry 1996: 49–50). However, forgetting, as indicated by de Certeau (1986: 3) 'is not something passive, a loss, but an action directed against the past'.

Since the prime function of memory 'is not to preserve the past but to adapt it as to enrich and manipulate the present' (Lowenthal 1985: 210), it is understandable that great political changes bring about changes of symbols. The outstanding exception in Ljubljana is the monument to Napoleon, a reminder of the 'bright era of the French occupation' (1809–1813). It was unveiled in 1929, when Slovenia was no longer a part of the Austro-Hungarian Empire, and can be seen as an embodiment of a continuing

Figure 9 Boris Kidrič (Source: Historical Archive of Ljubljana).

tension between immutable aspects of the past conserved in the present, in
contrast with the past as transformable and manipulable – yesterday's
villains as today's heroes (cf. Middleton and Edwards 1990: 8). On one side
of the Illyrian Column is a gilded bronze head representing Illyria and
Napoleon's head wearing a laurel crown is on the other. On the lower part
there are inscriptions on all sides, including verses by Valentin Vodnik,
(who, before the French occupation, constantly criticized Napoleon):

Napoleon says
Illyria stand up
a new spirit breathes into Slovenians
the spirit of Napoleon ...

On 21 June 1929, *Slovenski Narod* (The Slovenian Nation) explained why the Slovenians, and especially the people of Ljubljana, commemorated Napoleon and his Illyrian Provinces so late: 'The German government system ruled their thinking and their feelings. Since Germans hated France and Napoleon in particular, well-educated Slovenians remained insensible towards France, Illyria and Napoleon' (Platon 1929).

During the nineteenth and twentieth centuries, public monuments in Ljubljana were erected with the object of rendering present and visible that which was absent and invisible: Slovenian history. While public monuments in Ljubljana were time and again explicitly declared to be means for the commemoration of historical personalities and events, and elites were well aware of the power of remembrance, they were no less conscious of the supremacy of oblivion. Unable or unprepared to confront the monuments of past epochs, the protagonists of political changes regularly destroyed them. They saw public monuments as a mere tool with which to reconstruct the Slovenian past as a linear and straightforward story connecting the present with the mythical past. Using the past in the service of their future aims, they made use of history as a conflict between good and evil that contained a purpose that led it inevitably to an end (Plumb 1969: 98). In this scenario, today's heroes ('we are like them') created by political change not only repeatedly displaced the old ones as yesterday's villains ('we are not like them') but also exorcized the memories of what had hitherto been celebrated as a glorious age. When the First World War ended, all monuments in Ljubljana that were expressions of Austrian patriotism were removed. During the Second World War monuments that were expression of Yugoslav patriotism were demolished. Finally, after Slovenia became an independent republic the new authorities got rid of Tito's monuments. At the end of the day, no public monument testifies to Slovenia's historical links with Austria or Yugoslavia.

References

Certeau, M. de (1986) *Heterologies: Discourse on the Other*. Manchester: Manchester University Press.

Cesarjev (1916) 'Cesarjev 86. rojstni dan v Ljubljani. Odkritje cesarjevega spomenika na ljubljanskem Gradu.' *Slovenec*, 18 August 1916, p. 4.

Cleghorn, G. (1824) *Remarks on the Intended Restoration of the Parthenon of Athens as the National Monument of Scotland.* Edinburgh: Archibald Constable and London: Hurst, Robinson and Co.

Connerton, P. (1989) *How Societies Remember.* Cambridge: Cambridge University Press.

Čopič, Š., D. Prelovšek and S. Žitko (1991) *Outdoor Sculpture in Ljubljana.* Ljubljana: Državna založba Slovenije.

Delo.

Ferro, M. (1984) *The Use and Abuse of History, or, How the Past is Taught.* London: Routledge and Kegan Paul.

Firth, R. (1973) *Symbols: Public and Private.* London: Allen and Unwin.

Grazioli, E. (1941) *Notizie di Carattere Politico*; III verbale. L'anno 1941/XIX, il giorno 3 luglio, alle ore 10.30, nella sala delle riunioni dell'Alto Commissariato di Lubiana. Arhiv Republike Slovenije, Collection Visokega komisarja za Ljubljansko pokrajino, Fascicle 1: Seje konzult.

Gross, D. (2000) *Lost Time: On Remembering and Forgetting in Late Modern Culture.* Amherst: University of Massachusetts Press.

Hojda, Z. and J. Pokorný (1996) *Pomníky a zapomníky.* Praha and Litomyšl: Paseka.

Hribar, I. (1903) *Stavbnemu odseku in odseku za olepšavo mesta občinskega sveta v Ljubljani.* Manuscript in Zgodovinski arhiv Ljubljana, Reg. I., fascicle 2021 Spomenik Francu Jožefu, folia 397.

—— (1983/84) *Moji Spomini.* Ljubljana: Slovenska Matica.

Jeglič, A.B. (1908) *Dnevnik.* Nadškofijski Arhiv Ljubljana, Collection Škofijski arhiv Ljubljana, Diary of Bishop Jeglič 1908–1912, box 13/1c.

Kamnogorski, L.T. (1850) 'O Prešernovim spominku.' *Novice,* 10 April, p. 61.

—— (1858) 'O Vodnikovem godu.' *Novice,* 27 January, pp. 29–30.

Kos, J. (1997) *Glejte ga, to je naš Prešeren!* Ljubljana: Kiki Keram.

Kovačič, D. (1996) 'Del zgodovine sta Kardelj in Kidrič in ne njuna spomenika.' *Slovenec,* 13 April 1996, pp. 30–1.

Levec, F. (1888/89) 'Vodnikov spomenik v Ljubljani.' *Ljubljanski Zvon,* pp. 445, 765.

Lowenthal, D. (1985) *The Past is a Foreign Country.* Cambridge: Cambridge University Press.

—— (1996) *Possessed by the Past: The Heritage Crusade and the Spoils of History.* New York and London: The Free Press.

Mach, Z. (1993) *Symbols, Conflict, and Identity: Essays in Political Anthropology.* Albany: State University of New York Press.

Majar, M. (1858) 'O spomeniku Vodnikovem.' *Novice,* 6 October, pp. 316–17.

Mal, J. (1928) *Zgodovina Slovenskega Naroda.* Celje: Družba sv. Mohorja.

Maleš, M. (ed.) (1940) *Spomenik Viteškemu kralju Aleksandru I. Zedinitelju v Ljubljani.* Ljubljana: Odbor za postavitev spomenika Viteškemu kralju Aleksandru I. Zedinitelju v Ljubljani.

Mariborski odbor za spomenik Slomšku (1865) 'Poziv pomagati k stavbi spominka v čast pokojnemu knezu in vladiki lavantinskemu Antonu Martinu Slomšeku.' *Novice,* 6 December, p. 396.

Middleton, D. and D. Edwards (eds) (1990) *Collective Remembering.* London: Sage.

Mumford, L. (1938) *The Culture of Cities.* London: Martin Secker and Warburg.

Murko, M. (1891) 'Fr. Prešeren: Slavnostni govor dr. M. Murka pri Prešernovi devetdesetletnici na Dunaju, dne, 3. grudna 1890.' *Ljubljanski Zvon,* pp. 81–7.

Novak, J. (1858) 'O Vodnikovem spomeniku.' *Novice*, 10 November 1858, p. 356.

Novice.

Platon (1929) 'Napoleonov spomenik v Ljubljani.' *Slovenski narod*, 21 June 1929, p. 2.

Plumb, J.H. (1969). *The Death of the Past*. London: Macmillan.

Shackel, P.A. (2001) 'Public Memory and the Search for Power in American Historical Archaeology.' *American Anthropologist*, vol. 103, no. 3, pp. 655–70.

Shirer, W.L. (1988) *Berlin Diary: The Journal of a Foreign Correspondent 1931–1941*. Boston: Little, Brown and Company.

Slovenec.

Slovenski Narod.

Šnuderl, M. (1993) *Dnevnik 1941-1945*. Maribor: Založba Obzorja.

Stibrič, J. (1996) 'Zbiranje podpisov proti rušenju spomenikov.' *Delo*, 6 July 1996, p. 16.

Taylor, T. (Edited and Compiled) (1853) *Life of Benjamin Robert Haydon, Historical Painter: From his Autobiography and Journals*. London: Longman, Brown, Green, and Longmans.

Teski M.C. and J. Climo (eds) (1995) *The Labyrinth of Memory: Ethnographic Journeys*. Westport and London: Bergin and Garvey.

Trstenjak, D. (1858) 'Novoletnica.' *Novice*, 6 January 1858, pp. 1–2.

Tuma, H. (1937) *Iz mojega življenja*. Ljubljana: Naša založba.

Urry, J. (1996) 'How societies remember the past.' In: S. Macdonald and G. Fyfe, *Theorizing Museums: Representing Identity and Diversity in a Changing World*, pp. 45–65. Oxford and Cambridge, MA: Blackwell and The Sociological Review.

Žavčanin, K. (1858) 'O zadevi Vodnikovega spomenika.' *Novice*, 24 November 1858, p. 372.

Žolnir, N. (1996) 'Usoda spomenikov Kidriča in Kardelja skoraj odločena.' *Delo*, vol. 31 January 1996, p. 6.

Chapter 6

SIGN OF THE TIMES
Symbolic change around Indonesian independence

Freek Colombijn

Introduction[1]

Dutch colonial power transferred sovereignty to the independent Indonesian state on 27 December 1949. The municipal archive of the city of Medan does not contain a trace of this transfer except for an instruction about which flag to fly.[2] A flag is just a symbol and seemingly of little import. Yet, flying the correct flag was apparently the most important thing to be done on that day, which since then has been hailed as a turning point in every history book on Indonesia. The very first change in another city just liberated from colonial suppression, the East Timorese capital Dili after twenty-four years of Indonesian rule ended in 1999, was also symbolic. The East Timorese government substituted the former name of a main street, Avenida Almirante Américo Tomas for Avenida dos Direitos Humanos, Human Rights Avenue (Ramos-Horta, cited by Colombijn 2001: 5). At the time Dili was devastated,

---

[1] I am grateful to Martine Barwegen, who collected some of the data, Rosemary Robson for her English corrections, and the participants of the conference on urban symbolism (Leiden, 16–18 June 2004) who gave stimulating comments.

[2] All government offices and houses of high-ranking civil servants had to put out the red-and-white Indonesian flag. Offices of the governments of the state (*negara*) of East Sumatra (then a member of the federal United States of Indonesia) could add their own flag on the left side. Citizens were allowed, but not obliged, to put out a flag. Indonesian citizens put out only their own flag possibly with the *negara* flag. Foreigners were allowed to add their own flag, if the Indonesian flag hung on the right side and if both flags were of equal size and at equal height. It was forbidden to put out the foreign flag without the Indonesian one. Brief burgemeester Medan aan diverse ambtelijke hoofden, 17-12-1949, Arsip Bagian Umum Pemerintah Kota Medan (Arsip Medan), copyboek 5436/B-7.

and other things may have looked in more urgent need of attention than a name-board, but not in the view of the East Timorese government. Translucent bunting and road signs, with letters usually too small to be read by the persons seeking their way, are very important to some people, because of the power of their symbolic message. Therefore, they are worthy of being studied.

This article deals with the symbolic changes that took place against the background of a political process that culminated in the decolonization of Indonesia. The Dutch had begun to build up a territorial presence in the Archipelago in the early seventeenth century. By the early twentieth century the colonial state had subjugated almost all the land area that is nowadays known as Indonesia. While the Dutch were rounding off their empire in a military sense, political nationalism had begun to emerge. The colony was ineluctably drawn into the Second World War when the Japanese forces attacked and overran the Dutch colonial army. From March 1942 to August 1945 the Japanese ruled Indonesia, which they occupied mainly as a supplier of raw materials necessary to their war machine. On 17 August 1945, two days after the Japanese surrender to the Allied forces, two Indonesian nationalist leaders, Sukarno and Mohammad Hatta, proclaimed the Republic of Indonesia. The Dutch, who misjudged the situation totally, returned to their former homes to take up their old life again and tried in vain to crush the Indonesian Revolution by diplomatic and military means. The Dutch effectively administered most cities until the formal transfer of sovereignty in December 1949. The new country started as a federal state with the Republic proclaimed by Sukarno and Hatta, East Sumatra, and Pasoendan (West Java) as the biggest constituents states, but within a year all states had merged into one unitary Republic of Indonesia. Indonesia is nowadays the fourth most populous country in the world. The archipelago stretches over a distance equivalent to that from Ireland to the Urals and consists of about 13,000 islands.

By focusing on the decolonization process as the drive behind symbolic change, I have consciously restricted the analysis of symbolism to a political perspective. Much symbolism that does not refer to power relations, but, for instance, to life cycles (Firth 1973: 206), is not taken into account. Symbols from the sacred domain, such as holy graves, and symbols from the ethnic domain (such as Chinese graveyards), which were not contested until the 1970s, are likewise omitted. Symbols of modernization, such as the many offices built in the boomtown Medan between 1918 and 1924 (Loderichs *et al.* 1997: 32–3), the new airport of Surabaya (*Soerabaiasch-Handelsblad* 29-3-1940) or the Hotel Indonesia, the highest hotel in Southeast Asia when it was completed in 1962 (Bain 2002: 38), also fall outside the scope of this article.

Data were collected from published photographs, newspapers and other contemporary publications, the municipal archive of Medan, and historical studies. The analysis is hampered by its historical distance from the events it deals with, because ideally people should be asked which meaning they ascribe to a certain symbol. Insofar as I have interviewed people about what they thought of a certain symbol in the years 1930–1960, they can only answer in retrospect.

Ideally, a researcher studying the symbolism of one particular town should enter that town as a *tabula rasa* and allow the research to be led by what people say is significant (cf. Glaser and Strauss 1967). In Yogyakarta the university founded immediately after independence, Universitas Gadjah Mada (1949), may be the most important symbol of decolonization (Bambang Purwanto, pers. comm.) and in Surabaya it is perhaps the Jembatan Merah, the 'Red Bridge', which was contested in a fierce battle. Perhaps the greatest of all symbolic changes was that of the name of the capital: from Batavia to Jakarta. This article, however, makes the beginning of a comparative analysis between different towns of Indonesia, and if we do not want to compare apples and oranges, I must determine beforehand which types of symbol carriers (Nas 1993a: 15) I shall focus on.

This analysis of the symbolic changes induced by political conflict between 1930 and 1960 is structured around three types of symbol carriers: statues (petrified images), street ceremonies such as parades or putting out a flag (live images), and street names (worded images). By the focus on one type at a time, we are better able to see the changes in each type in the course of political events. The choice for these three types of symbol carriers – statues, ceremonies, and street names – makes an assessment of the differential power of material, behavioural and discursive signifiers possible (Nas, Jaffe and Samuels, this volume).

A secondary aim of this article is to rethink the study of urban symbolism as a subdiscipline of urban anthropology, and it is to this aim that I turn first.

The study of urban symbols
The volume *Urban Symbolism*, edited by Peter J.M. Nas (1993b), marked the start of a decade of enthusiastic studies. The book was not the first to record the study of urban symbolism (for Asia, see for example Heine-Geldern 1942; Jeffrey 1980; Stutterheim 1948), but it achieved seminal status because the editor urged his network of colleagues and students to take up the topic (for example Jezernik 1999; Nas 1998a, 1998b; Taal 2002).

Few persons have embraced Nas' term 'symbolic ecology' for the study of urban symbolism (Nas 1990, 1993a),[3] but his tool kit of nine concepts and the idea (to look at urban symbols) as such have proved to be very stimulating. About sixty articles have appeared which are directly related to Nas' suggestion to focus on urban symbols; some of them give a theoretical elaboration of the original nine concepts.

After a decade, we have learnt a considerable amount about the symbolism of individual towns and cities. Most of these studies are mainly descriptive (which is, in my opinion and contrary to the common view, not an inferior kind of research). The articles and chapters are studies of one city and the authors have studied symbols in order to better understand the city concerned. Few have reflected on the nature of symbolic meaning itself. Building on this base, I see three directions in which to further our understanding of urban symbolism.

The first direction is a comparison of cases, which will generate several theoretical questions, and perhaps some new answers. One question is: when do people create 'referential symbols' (Nas and Van Bakel 1999: 188)? For example, the national monument in Kuala Lumpur (soldiers planting the Malaysian flag) is a conscious quotation of the central statue in Arlington National Cemetery in Washington (American soldiers planting the Stars and Stripes). Both statues are the work of the same sculptor, Felix de Weldon, and the similarity is reinforced by the fact that the real Malaysian flag, which flies above the bronze figures, also has stars and stripes.[4] We can compare the Malaysian monument with a small monument at the military cemetery in Pekanbaru (the capital of an Indonesian province), which is a short and stocky replica of the national monument in Jakarta. Are such referential symbols a sign of ambition to rise to the same level, or recognition that one will remain secondary forever? I believe that a city that wants to build an image cannot refer only to other cities.

Another question generated by comparative analysis concerns the degree of the uniformity of the symbolic message. It seems that symbols of political leadership are altered after a political change. Political symbols are dynamic, but at any moment in time, the message is uniform. For example, the statue

---

[3] The term refers to the human ecology of the Chicago School, but is, I believe, inappropriate, because urban symbolism as practised by Nas lacks the ecological analogy of invasion, succession, and competition between different species. At first Nas (1990, 1993a: 15) used the term symbolic ecology for the spatial distribution of symbols, which made more sense, and only later has the term referred to the study of symbolism in general.

[4] The statue at Arlington itself refers to a photograph taken during World War Two on the Pacific island of Iwo Jima.

of Saddam Hussein in the centre of Baghdad was knocked off its pedestal after Saddam Hussein had clearly been defeated in the Second Iraq War. The Berlin Wall was stormed by the public only after it was clear that the East German and Soviet Union governments were no longer able or willing to defend it. Mao Zedong's image was gradually removed from the streets in China when his successors set about dismantling his political and ideological legacy to an increasing degree.[5]

I hypothesize that the uniform message of political symbols contrasts with the pluriform message of ethnic-religious domains. The symbol is not changed after one party has won a decisive victory, but the symbol itself becomes the stake in the conflict. Contested Chinese cemeteries are an example of an ethnic symbolic conflict. This, at least, was the case in a prolonged conflict between Chinese residents and a non-Chinese municipal government about the closure of the Chinese cemetery in Padang (Colombijn 1994: 337–41). A similar conflict about the Chinese cemetery in Singapore arose between local residents and proponents of modernization (Yeoh 1991). Houses of prayer which come under attack are an example of a symbolic conflict of a religious nature. In 1999, an armed conflict broke out along Christian–Muslim lines in Ambon (Indonesia) between groups who had hitherto been friendly neighbours. The violence in Ambon City, which continues to simmer up to the present, has focused on a centrally located mosque and church. One result of the fights has been religious residential segregation along a line that has been dubbed 'Gaza Street' (another referential symbol). The Ambon City situation can be compared to the fights between Hindus and Muslims in Ayodhya contesting a sacred site where both a Hindu temple and mosque were built and demolished in the past (Indurthy 1994; Panikkar 1993). The violence in both Ambon City and Ayodhya may be fuelled at certain times by electoral campaigns in which politicians do not scruple to play on religious sentiments.

Yet another theoretical question is why in the wake of decolonization do some new regimes destroy symbols of their colonial predecessor (Indonesian examples will be given below) and why others do not. When Singapore was at the point of being decolonized and about to join the Malaysian Federation in 1963, its future looked bleak. Prime Minister Lee Kuan Yew recalled how a Dutch consultant and one of the main architects of his city-state's economic success, Albert Winsemius, gave two pieces of advice on his first

---

[5] He was not, however, totally removed from the scene. His mausoleum still occupies a central place in Tian'anmen Square and a giant portrait of Mao decorated the gate to the Forbidden City. In 1992, I also observed a statue of Mao in the city of Chengdu.

visit in 1961: first, to eliminate the communists, and, second, not to remove the statue of Thomas Stamford Raffles, the British founder of Singapore. Lee continues: 'To keep Raffles' statue was easy. My colleagues and I had no desire to rewrite the past and perpetuate ourselves by renaming streets or buildings ... Letting it remain would be a symbol of public acceptance of the British heritage and could have a positive effect [on investors]' (Lee 2000: 50).

More theoretical questions can be approached from a comparative perspective. When natural features have a large impact on the symbolism of a city, as is the case in Cape Town (Nas, te Velde and Samuels, this volume) or Canberra, is this because the landscape is so overpowering or because there is a need to suppress human (conflicting) features? Does a relationship exist between location (unknown; one permanent point; moving around or temporarily at one point; multiplied at several points; or embracing the whole town) and power of a symbol? Why do people become saturated with some symbols to the point they no longer see the symbol? When are forms of counter-symbolism or symbolic resistance, such as graffiti, the harbinger of a revolution and when are they a safety valve to let steam off and canalize resistance in harmless directions?

A second direction in which the study of urban symbolism can be developed is to return to an emic approach. Too often, studies are based on 'anthropological tourism' (Colombijn 1998: 566–8). At worst, the study reflects what the researcher sees in the symbol. At best, the study reproduces the intentions of the creator of a symbol. Both miss the point that the essence of a symbol lies in what people think of it. A symbol is a relationship between an object (statue, flag, building, song, coat of arms, street name) and a meaning outside the object. What matters is not the object, but the meaning that people give to the object. The meaning given to the symbol by both the creator[6] and the receiver of the message is what counts.

The creator and receiver of the symbolic message can have very different ideas about the meaning of the message. The Dutch novelist Renate Dorrestein (2000: 191) complained that literary critics persistently sought for symbolic meaning in the names of her protagonists, where there was no such meaning. For example, a person called Agnes Stam allegedly referred to Agnus Dei (the Lamb of God). Ernest Hemingway (quoted in Phillips 1984: 4) commented on the *The Old Man and the Sea*: 'There isn't any symbolysm [sic]. The sea is the sea. The old man is an old man. The boy is a boy and the

---

[6] Perhaps we must make a distinction between the authority that ordered the symbol and the creative person who gave the symbol its form.

fish is a fish. The shark[s] are all sharks no better and no worse. All the symbolism that people say is shit.' Understanding the writers' irritation, I think that from an anthropological perspective the symbolism given to words and objects by the audience counts too and cannot be relegated to the realm of faeces. It is the reader who picks up a book and turns a text into a message.

We can safely assume that also within the audience, different people give different meanings to the same object. This variation of meanings must be studied, and cannot be reduced to the intentions of a creator of a symbol. This is especially the case where a symbol is not deliberately created (such as a statue or street name), but has evolved gradually. Symbols evolve gradually when new meanings are attached to natural features of the landscape or where old human-made objects (streets, squares, buildings) acquire a new meaning in the wake of a certain event. An example is the Abu-Ghraib prison in Baghdad. It was once built as a functional building, a prison, which later acquired strong symbolical overtones. First, it became a symbol of Saddam Hussein's terror. Then, in 2004, it became a symbol of American arrogance (or anti-Muslim hatred, racism, sexism, or whatever) after photographs were published of Iraqi soldiers being humiliated by American guards. A research methodology should take account of the variety of meanings and measure this variety (by interviews, or expression of public feelings in media). A good example of this kind of approach is a study of architects in Aceh by Peter Nas (2003). Although Nas did not study what the 'common people', whoever they may be, think of local architecture, he did interview five generations of architects and recorded how they assessed each other's work and how they voiced opinions about the symbolic quality of the designs. An interesting experiment in emic analysis is the study of mental maps of residents of Denpasar, Jakarta and Bukittinggi based on sketches drawn by informants (Nas and Sluis 2002). Given the fact that the latter study is based on small numbers of respondents, almost all students, the question of to what extent the results can be generalized is pressing.

The last remark to be made about the emic approach is that sometimes local people can have a different view of what constitutes a symbol from that of the researcher. Robin Jeffrey (1980: 485) has remarked for India that before the British colonization no statues of mortals were made. The idea of a statue of somebody who did not have supernatural powers was foreign. In Laos, where the communist party leader Kaysone Phomvihane tried to build a personality cult for over a quarter of a century, statues were also erected only to deities. Buddhas were 'stylized so as not to look like mortals' (Evans 1998: 114). Evans believes that the same problem exists throughout

Southeast Asia. Indeed, in Padang I observed that people were saturated with the symbolism and no longer noticed the statues (Colombijn 1994: 336, 350); this raises the question of whether the statues must be called symbols at all.

The third direction in which to further our understanding is to focus on processes of how single symbols are created. An excellent example is provided by Frank Pieke (1993), who studied the students' occupation of the Tian'anmen Square in Beijing in 1989. After having given an overview of the hegemonic (state) view of the square, he shows how students challenged the discourse of the Chinese Communist Party. Move and countermove of party leadership and students culminated in a symbolic pinnacle when the students erected the statue of the Goddess of Democracy, a replica of the Statue of Liberty. It was merely a 'referential symbol', of course not as powerful as the original, but in this context the Goddess was a very powerful protest, precisely because it referred to the allegedly Free West. The statue was deliberately placed outside the old, sacred north–south axis of the Tian'anmen square. Pieke (1993: 166) concludes that 'it is likely that this disregard for the cosmology of [… the Goddess of Democracy's] surroundings was intended to stress its iconoclasm, its rejection of traditional and communist China and its embrace of a foreign ideology'. The occupation ended with the massacre of hundreds, maybe thousands, of protestors, so that the square has acquired a new symbolic meaning, at least in the eyes of some people, namely the symbol of state suppression in China. For their part, the authorities were careful to kill protestors only in the side streets and denied, and were probably right, that blood had been spilled on the sacrosanct square itself.

This processual approach gives abundant insight into what a symbol is, and how different groups perceive a symbol. Moreover, studies of social processes potentially make the most exciting reading (Jaffe this volume; Klaufus this volume).

Keeping these three directions in mind, I can now specify the aim of this paper. In this paper I make a very modest attempt to work along the lines of my criticism. Of the three directions for future research – comparative, theoretical analysis; emic analyses taking account of variation between different groups of people; and processual case studies – the first is perhaps the most innovative. By comparing symbolic change around independence in different Indonesian cities, I hope to explain symbolic variation between cities. At the same time, I hope to show something of the variation of meanings which people give to symbol carriers and something of the processes by which new symbols were created.

Statues

Statues form a good starting point for the analysis because they are the only objects of the built-up environment with a purely symbolic function. An almost comprehensive list of colonial statues exists for the West Sumatran town of Padang (Colombijn 1994: 333–47). The Padang statues commemorated: Lieutenant A.T. Raaff (the first Dutchman to fight in the hinterland of Padang; the statue was erected shortly after his death in 1824); General A.V. Michiels (subjugator of West Sumatra, unveiled in 1855); W.H. de Greve (discoverer of coal deposits, unveiled in 1880); the Sarekat Adat Alam Minangkabau (an organization of conservative pro-Dutch Indonesians, unveiled in 1919); the Jong Sumatranen Bond (a moderately nationalist movement, inaugurated in 1920); W.A.C. Whitlau (a local administrator, the clock tower dedicated to him was built in 1922 and was already demolished in 1934 because it was considered a very ugly eyesore and an obstacle to traffic); and Soetan Masa Boemi (an Indonesian leader loyal to the Dutch cause, erected in 1930).

As a group, the statues in Padang form an interesting pattern. The Dutch army dominated the symbolic space. The most powerful statue was the Michielsmonument, a ten-metre high iron tower in the centre of the Michielsplein (Michiels Square). Size, form (a massive, neo-Gothic tower referring to European architecture) and site (in the second largest square) lent this statue its power (Figure 1). The fact that Dutchmen at least mistook it for a statue commemorating the Dutch admiral Michiel de Ruyter did not alter its political meaning of Dutch military dominance. In the biggest square, Plein van Rome, was found the other military monument, the Raaff needle. It had been moved to this site in 1913 because its original site on the coast was being eaten away by the force of the waves. Despite its size, the Plein van Rome was symbolically speaking less important than the Michielsplein. The Plein van Rome was used for such worldly activities as military exercises, football (forbidden on Michielsplein), and a market. I do not know of any photograph showing the Raaff needle and it must have been unimposing.

A prominent position of colonial soldiers is what one expects to find in colonial society. The less obvious feature of the Padang pattern is that the symbolism of the Dutch colonial military or state was not hegemonic. After 1855, almost during the full last century of Dutch colonialism, no statue with a military meaning was added. The funding of the statues forms the explanation why colonial might was not hegemonic. All statues were financed by sums collected from the public. The result was a pluriform symbolic message.[7] A significant detail is that all statues were planned and executed after the person had died or retired or after the event had occurred.

Figure 1 The Michielsmonument in Padang.

The same pattern of fundraising among the population can be found in other cities. In Surabaya, for example, the Chinese community presented a bust of the late mayor, Dijkerman, to the municipal administration.[8] In Bandung, to give another example, a monument to Dr Ir C.J. de Groot was

---

[7] This raises the question of how can one express oneself symbolically if one does not have money to erect a statue.

[8] Dijkerman was commemorated for being impartial in inter-ethnic relations, with the result that Chinese people became interested in seeking a seat on the city council and one Chinese became alderman (*Soerabaiasch-Handelsblad* 29-1-1930).

unveiled in January 1930 (*Deli Courant* 27-1-1930; *Soerabaiasch-Handelsblad* 25-1-1930).[9] For one Minahasa statue, funds were raised all over the archipelago (*Deli Courant* 30-1-1930) in a way Anderson (1992) calls 'long-distance nationalism'.

To test the idea that the state or military is not hegemonic, we need a more comprehensive picture of statues than we have of Surabaya or Bandung. Medan is a city about which we are better informed. Here, the pattern already found in Padang – of initiatives for new statues coming from the people, with a pluriform symbolic message as result – was also encountered. In Medan, the well-to-do Chinese merchant Tjong A Fie paid for the clock tower on the town hall, built in 1912 (Loderichs *et al.* 1997: 16). The beautiful Virtuous Bridge (Jembatan Chen Tek) over the River Babura (built 1916) is also attributed to Tjong, although the name of one of his Chinese relatives is inscribed on the bridge. A fountain was dedicated to Jacob Nienhuys, the pioneer who started tobacco cultivation in north Sumatra, in 1915.[10] In 1924 the Deli Planters Association (Deli Planters' Vereeniging) erected a statue of J.Th. Cremer in front of their new office building (Loderichs *et al.* 1997: 33, 112).[11] The year 1924 was also the year that the first flight was made from the Netherlands to Indonesia. The plane touched down on Indonesian soil on the race course in Medan. The following year a memorial, crowned by a model of the plane, was constructed (Loderichs 1997: 148–9). The main war memorial, the Tamiang Monument, erected to celebrate a military campaign in Aceh in 1893, was much older, dating from the nineteenth century (Loderichs *et al.* 1997: 21).

What remained of the pluriform message of the statues after the Japanese forces occupied the Netherlands East Indies? I have argued elsewhere that in Padang the Japanese destroyed all Dutch statues (the Michielsmonument, which was allegedly remoulded into bullets, went first) and left only the Jong Sumatranen Bond statue in peace. They did not build monuments themselves (Colombijn 1994: 347). In Jakarta the Japanese also demolished the monuments but 'were not in power long enough to erect such symbols themselves' (Nas 1993a: 24; see also Dinas Museum dan Sejarah 1980). The

---

[9] De Groot was a pioneer in radio technology, or, in his own words, a '*radioot*'. The first wireless telegraph office in the Netherlands Indies (on Sabang in 1911) was constructed by him. In 1923 he established, from the tropical side, a radio connection with the Netherlands. He died in 1927, but did not have a specific relationship with Bandung (Visser 1979).

[10] The present fountain on this location is, contrary to the idea of Suti (1979: 22), not the same as the original one.

[11] I do not know when it was removed from the street, but much to my surprise I found it inside a municipal office building in 2003.

negative historical image of the Japanese as a destructive, but not a constructive, force does not apply throughout the whole of Indonesia.

More of the pluriform symbolism survived the Second World War than can be deduced from Padang and Jakarta. Nas himself remarks that the Amsterdamse Poort (Amsterdam Gate) in Jakarta, which had become a monument (but a symbol for what?) by the Second World War was spared by the Japanese (Nas 1993a: 24). The Van Heutszmonument in Batavia[12] was damaged during the Japanese period and again during the Indonesian Revolution, but not destroyed. It was not demolished until the 1960s (Dinas Museum dan Sejarah 1980: 5; Voskuil 1993: 129). According to two elderly residents of Medan, both with a sharp memory, Dutch statues were not removed during the Second World War or in the Revolution, but fell victim to fanatical nationalists somewhere between 1952 and 1956. The main war damage to a symbolic object in Medan was the requisition of the bell from the Reformed Protestant Church, which was claimed by the Dutch colonial air-raid defence. The bell was placed back in the church tower in 1948.[13]

The claim that the Japanese did not build monuments themselves is not valid for the whole of Indonesia. The famous Villa Isola in Bandung, for example, was converted into a museum commemorating the Japanese advance in Java (Voskuil *et al.* 1996: 74). Loderichs *et al.* (1997: 61) publish a photograph of a Japanese statue in Medan. The Japanese army erected this tall, square pillar in the central square of Medan. It was inaugurated in December 1944 to commemorate the Japanese victory in Sumatra. I have not found other evidence, written nor oral, about this monument, but the photograph is unmistakable proof. The statue raises the question of whether there may not have been more monuments, which were not fortunate enough to be captured in a chance photograph.[14] The archetypical material monument in Japanese culture, however, is not the statue but the shrine (Nelson 2003). I do not know of shrines for Japanese soldiers slain in Indonesia, like those made by relatives in Papua New Guinea (Petr Skalník, pers. comm.).

---

[12] The Van Heutszmonument was built between 1926 and 1933. Van Heutsz was victor in the Dutch–Aceh War (1873–1904); he died in 1924.

[13] Afschrift besluit Secretaris van Staat, Hoofd Departement van Binnenlandsche Zaken 20-7-1948, Arsip Medan, copyboek B13/2/13.

[14] Another example is a photograph of a statue commemorating the victims fallen during the construction of the Thailand–Burma railway; the Japanese erected it at the Tamarkan railway bridge (Thailand) in 1942 or 1943. The statue has the same robust shape of the Medan statue. Koninklijk Instituut voor Taal-, Land- en Volkenkunde (Leiden), Archive and Image, photograph 25531.

During the interlude between the Japanese surrender and the definitive Dutch retreat from power, the Dutch, believing they would stay for decades, resumed erecting statues. A Dutch battalion erected a small pentagonal statue to honour its fallen on the *alun-alun* (central square of a Javanese town) of Malang. After the transfer of sovereignty the municipal administration ordered the demolition of the statue (van Schaik 1996: 80). A monument in Manado commemorated Indonesian and Dutch victims of the Japanese occupation. Dutch rule ended before the statue was completed (Lapian 1999). With the wisdom of hindsight, these two statues proved to be entirely ephemeral,[15] therefore it is better to return to 1945 and the events on the Indonesian side.

The proclamation of Independence and the ensuing Revolution marked a watershed in the history of Indonesia. Intuitively, one expects that this turning point would have been marked by great symbols, especially because the first president, Sukarno, was a man of oratorical skills, fond of great words, grand gestures and outward appearance in general. But also, regardless of the personality of the first president, it seems likely that revolutionaries would have wanted to communicate their message to the public in symbols. To what extent was this expectation of lively symbolic activity in general and the personal involvement of President Sukarno in particular correct?

During or immediately after the Indonesian Revolution, urban administrators began to think of liberation monuments. One of the first monuments was the Proclamation Monument (Tugu Proklamasi) erected in front of Sukarno's house-cum-office, exactly where he and Mohammad Hatta had declared Independence. It was a short obelisk inaugurated on the first anniversary of the proclamation, 17 August 1946, when Jakarta was under control of British forces and Sukarno himself had withdrawn from Jakarta. Sutan Sjahrir, Prime Minister and Minister of Foreign Affairs of the Republic, who had remained in Jakarta to negotiate with the Dutch, presided over the inauguration ceremony. British troops prevented adherents of the Republic approaching the obelisk during the inauguration (Leclerc 1993: 40).

Sukarno's office was given to the Ministry of Education and Culture to be preserved as a historic building in 1951. However, the building fell into disgrace, because Sutan Sjahrir had used Sukarno's office as a base when he

---

[15] The statue in Manado still exists. It was grabbed by the adjacent church, which added a cross to one side, so that the monument has lost its former meaning.

negotiated the Linggajati Agreement (1946).[16] As this agreement was gradually seen as a betrayal of Republican ideals, the office shared in this disapprobation. The Tugu Proklamasi was dubbed Tugu Linggajati; the nickname was erroneous but fatal. By 1961, Sukarno himself ordered the demolition of the monument together with the house. The main error of the Proclamation Monument, according to Jacques Leclerc, was that it was inaugurated by Sjahrir instead of by Sukarno himself, but it also lacked grandeur. In 1972, after the fall of Sukarno, a duplicate was put in its place (Dinas Museum dan Pemugaran 1999/2000: 11–12; Leclerc 1993: 41).

Nowadays, Jakarta is associated with a number of grand statues in Soviet realist style, which are attributed to Sukarno (Nas 1993a). It is noteworthy that all these statues were conceived after President Sukarno's soft coup d'état of 1959, when he obtained wide-ranging executive powers and began to interfere actively in the capital's administration.[17] As long as the Jakarta administration preserved a distance from the national government, until 1959, there was apparently no strong incentive to create symbols. It is also worth noting that four statues were wholly or partly financed by somebody or some instance other than the government: one was a gift from the Soviet Union and another from an Italian Consul-General; until the completion of the first phase in 1968 the National Monument (Figure 2) was paid by donations from the public; a fourth statue was partly paid by Sukarno himself (he allegedly sold a private car to raise the money), because he feared it would remain unfinished (Dinas Museum dan Pemugran 1999/2000; Dinas Museum dan Sejarah 1980). These four statues partly financed from non-state sources show that even at the height of state

---

[16] The Linggajati Agreement was the last attempt to solve the Indonesian–Dutch conflict peacefully before large-scale fighting broke out. The agreement met with much opposition from both the Indonesian and Dutch politicians at home, because both sides believed too many concessions had been made to the other side. Therefore, hawks on both sides disparaged the Linggajati Agreement as soft.

[17] The idea of erecting a national monument on Medan Merdeka (Independence Square) was first conceived in 1955, and gained momentum after 1959. The eventual 1961 design, drawn by Soedarsono, was according to the artist an elaboration of an idea sketched by Sukarno himself (Leclerc 1993: 41–4). Kwee Hin Goan (pers. comm.) believes that the design was actually strongly influenced by two plans drawn by architect Silaban and a group of students from the Technical Institute of Bandung (including Kwee himself). The National Monument was opened to the public in 1972, so that Sukarno's successor had the chance to claim Monas as partly his work. Many persons apparently have an interest in claiming its authorship. Another statue in Soviet style, the Young People Development Statue (Patung Pemuda Membangun, nicknamed the 'Pizzaman'), was constructed in 1971 and belongs to the post-Sukarno times.

monument building, outsiders still undermined the state hegemony.

The Proclamation Monument in Jakarta was not the oldest resistance monument in Indonesia, but other initiatives were delayed by wartime activities. As early as 28 October 1945, the foundation stone was laid for the Statue for the Young Ones (Tugu Muda) in the central open space (*alun-alun*) of Semarang.[18] The next month, however, fights with both Japanese and Allied forces drove the Republican government out of Semarang. In 1951, the urban administration took the initiative to erect the statue after all, but chose a different location, at the Freedom Park (Taman Merdeka). The statue was financed mainly by public donations and had the form of a needle with a flower in the top. It was inaugurated in 1953 (Budiman 1979: 59–60).

The Republican city administration of Malang had immediate plans to erect a statue in front of the town hall. On 17 August 1946, the first anniversary of the Proclamation of Independence, the foundation was laid.

The design consisted of four jointed stone pillars rising upwards, which perhaps visualized the famous bamboo spears (*bambu rucing*) used by young revolutionaries. The monument was almost ready when Dutch forces took the city in 1947. In August 1948 the statue was embellished by a Dutch flag and a crown, on the occasion of the celebration of the fiftieth anniversary of Queen Wilhelmina's reign. Later, the whole statue was shattered and only the pedestal remained. This was an invitation to put something different on the empty pedestal. In 1950, a giant doll was made, representing the exhausted Indonesian people. Finally, the statue was rebuilt as it had originally been imagined and was inaugurated by President Sukarno in 1953 (van Schaik 1996: 122).

Other city administrations only began to think of a monument after the struggle was over. The urban administration of Padang built a freedom monument within three months of the transfer of sovereignty. The statue was erected on exactly the same spot as the former Michielsmonument and the location underscored that the era of Dutch colonialism had gone for good (Colombijn 1994: 348). In Surabaya the Heroes' Monument (Tugu Pahlawan) was erected in front of the governor's office, on the site where the former Court of Justice had stood. It commemorates the fighting in November 1945, which was the bloodiest hostility in any city during the Revolution. In Medan, a foundation to raise funds for a so-called National Building Medan (Gedung Nasional Medan) was established in 1951. The foundation asked the urban administration to levy 25 cents on each cinema

---

[18] This was only two weeks after fights between *pemuda* and Japanese soldiers had raged (Frederick 1999: 26)!

ticket to help pay for the monument.[19] A Statue Building (Gedung Arca) was inaugurated in a Statue Park in 1956. A poor-quality photograph does not reveal many details, but a small statue is discernible in front of the Gedung Arca (*Madjalah Kota Medan* 3, 25, August 1956: 7–9). Apparently the Gedung Arca was not satisfactory. In 1959, the city council decided to erect a Statue for Fighters from the Medan Area (Tugu Pedjuang Medan Area). The fundraising committee asked the municipal council to levy 10 per cent on all permits issued to pay for the statue.[20]

The examples from Jakarta, Semarang, Malang, Padang, Surabaya and Medan show that it took time and money to create a monument. The young republican urban administrators lacked the resources to build dominant monuments straightaway. A well-chosen location, in Padang and Surabaya for instance, could help strengthen the message that the colonial times had gone forever. Live urban symbols, such as parades and putting out a flag, are in theory much easier, cheaper and quicker to organize. Therefore, it is possible that these were more important during the short rule of the Japanese and in the first years of Indonesian independence.

Street ceremonies: parades and putting out the flag
In colonial times events related to the Dutch royal House of Orange (Queen's Birthday, the fortieth anniversary of Queen Wilhelmina's reign, the marriage of Crown-Princess Juliana to Prince Bernhard, birth of a heir) were celebrated by parades and other ceremonies. Temporary ceremonial gateways, arches, and flag towers were erected and precisely because they were built for the occasion, they drew attention. The Arab, Chinese, and (pre-war) Japanese communities also built such gateways to express the allegiance of their community to the Dutch overlords. As these gateways reflect in Oriental details their Arab, Chinese, or Japanese background, they were also a symbol of multiculturalism (see for example the photographs in Broeshart *et al.* 1994: 19, 77, 91, 148; Loderichs *et al.* 1997: 42–3, 46; Van Schaik 1996: 113). The practice was resumed after the Japanese period wherever the Dutch regained control of the cities, as if no Indonesian Revolution was taking place. For example, the Indian residents of Malang,

---

[19] Letter Djaidin Poerba to Kementerian Perekonomian, Kantor Pengendalian Harga in Djakarta, 20-11-1951, Arsip Medan, copyboek 6655/BC-2.
[20] Letter Ketua dan Sekretaris Panitia Tugu Pedjuang Medan to Ketua DPD Peralihan Kotapradja Medan 24-10-1959, Arsip Medan.

Figure 2 The National Monument under construction, 1964 (Pelaksana Pembina Tugu Nasional 1978: 64).

mainly originating from Bombay, erected a statue on the fiftieth anniversary of Queen Wilhelmina's reign and again on the installation of Queen Juliana (6 September 1948) (Van Schaik 1996: 82).

In the eyes of the Japanese colonizers, showing the flag was a quick and cheap symbolic expression of their newly acquired dominance. A photograph taken in Malang on the day of their entry, or shortly thereafter, already shows Japanese flags put out by the shops (Figure 3). Among the first Japanese regulations was a prohibition on flying the Dutch flag or the Indonesian nationalist flag, along with a prescription on how to hang out the Japanese flag (van Schaik 1996: 48–50; Voskuil *et al.* 1996: 73).

From now on, the birthday of the Japanese Emperor (29 April) and other feasts would be celebrated with parades and, among other attributes, a no doubt lukewarm floral tribute, and solemn bows from the residents (Raben

1999: 12; van Schaik 1996: 49, 57). On the first opportunity, April 1942, one month after the Japanese conquest, emblems of Dutch power, such as books, royal portraits, flags, and uniforms were burnt on a bonfire in Bandung. Three months later, the Chinese residents in Bandung were forced to burn portraits of Chiang Kai-shek and other symbols of the Republic of China publicly (Voskuil *et al.* 1996: 70–1).

The ceremonies provided opportunities for dangerous mockery and contra-symbolism. Sabotaging a ceremony was a form of 'everyday resistance' (Scott 1985), but it could be punished harshly. Three Dutchmen who were foolish enough to broadcast the Dutch national anthem for two weeks after the Dutch capitulation on the Bandung radio were beheaded (Voskuil *et al.* 1996: 72). Others fared better. Indonesian pupils mocked the obligatory bow (Maier 1999: 100). The Sultan of Serdang (near Medan) hung a photograph of the Japanese Emperor behind him when Japanese forces were entering his palace. Seeing the photograph, all soldiers made a bow to the Emperor, and hence to the Sultan (Tuanku Luckman Sinar Basarshah II, Sultan of Serdang, pers. comm.).

Figure 3 Petjinanstreet in Malang on 9 March 1942 or shortly thereafter (Courtesy of Netherlands Institute for War Documentation, NIOD).

When the Indonesian Revolution broke out, putting out a flag was often the quickest form of symbolic adherence. In many cities it has been recorded where and when the first Indonesian flag was raised.[21] By far the most famous 'first flag' was raised in Surabaya. According to the conditions of the Japanese surrender to the Allied Forces, the Japanese were still responsible for law and order and the Japanese authorities in Surabaya forbade the putting out of the Dutch flag. Nevertheless, on 19 September the first returned Dutch soldiers hung out the Dutch flag (horizontal red, white and blue bars) on the Yamoto Hotel (formerly known as Oranje Hotel, after the Dutch House of Orange). Indonesian youngsters, provoked by this gesture, climbed the hotel and tore of the blue strip, changing the flag into the red-and-white Indonesian banner (Figure 4). Young Dutchmen, who were present, were incited in their turn. In the ensuing fight several men died. On 1 October, Indonesian revolutionaries called on the public to lower the Japanese flags and replace them by Indonesian ones. This call triggered fierce fights, in which the Japanese strongholds in town were captured, most Japanese soldiers were interned and the revolutionaries gained control of huge amounts of weapons (Broeshart *et al.* 1994: 48–9; Dick 2002: 80; Han Bing Siong 2003). Not withstanding the importance of the spoils (including 19,000 rifles, 1,000 machine guns, 16 tanks, 62 armoured cars, and 2,000 trucks), the initial skirmishes were all about lowering or raising flags.

British allied forces took Surabaya after heavy fighting, which started on 10 November 1945 and continued until 1 December. Between September and December at least 15,000 Indonesian fighters and civilians, and about 500 British soldiers, 400 Dutch civilians, and 300 Chinese civilians were killed. Nowadays, 10 November is commemorated as Hari Pahlawan (Heroes' Day) throughout Indonesia. The celebrations start on the evening before, historically not entirely correct, with an imitation of the storming of the still extant Oranje Hotel. A white flag substitutes the Dutch flag at these celebrations, in order to prevent a diplomatic row (Broeshart *et al.* 1994: 53–6, 135; Dick 2002: 82–4). The conscious violation of the historical facts points all the more to the fact that a symbolic event (tearing off a blue strip) is more memorable than the loss of thousands of lives and the demolition of the city centre.

Ceremonies can make people feel ill at ease when form and content do not fit well. When the Dutch gradually retook Indonesian towns between 1945 and 1948 they often installed an Indonesian administrator to win the hearts and minds of the indigenous population. This policy resulted in uncomfortable situations. For example, the President (Wali Negara) of the

---

[21] For example, in Padang on 29 August 1945 and in Palembang on 3 October 1945 (Colombijn 1994: 348; Taal 2003: 72–3).

Figure 4 Indonesians climbing up to the Dutch flag, Surabaya, 19 September 1945
(Photograph by Abdoel Wahid; courtesy of G.H. Dekker, Broeshart *et al.* 1994: 48).

state Pasoendan, one of the constituents of the federal state of Indonesia in the making, took the salute on Queen Juliana's birthday, 30 April 1949 (Voskuil *et al.* 1996: 101). By then the Pasoendan leaders were already displeased with the Dutch refusal to grant real autonomy to their state. Likewise, the Indonesian mayor of Medan, Djaidin Poerba, gave a speech when corpses of Dutch soldiers were reburied in the just opened Dutch war cemetery. It was an odd speech, because topic and audience implied that the speaker should be a Dutchman.[22]

---

[22] Rede burgemeester Djaidin Poerba 23-11-1949, attached to letter Gemeente-secretaris

During the first years of independence the annual Proclamation Day would remain an occasion for public ceremonies drawing a large crowd spontaneously. For example, on the eleventh anniversary of the proclamation, tens of thousands of people gathered in Independence Square (Lapang Merdeka) in Medan. From there, a long procession walked to the War Cemetery. Later, in the heat of the day, thousands of schoolchildren sang a tribute to the governor. The previous day, several temporary gateways had been erected at the main roads and in urban kampongs. Multi-coloured lamps were hung up in front of shops and other buildings (*Madjalah Kota Medan* 3, 25, August 1956: 7–9). The state did not have a monopoly on ceremonies and Independence Day was not the only occasion to go out on the street. The Indonesian Communist Party, PKI, organized parades on Labour Day, 1 May (van Schaik 1996: 86).

Street names

The establishment of official street names 'introduces order and differentiation into an originally amorphous landscape' (Yeoh 1996: 221). Sandra Taal (2003: 154–5) has postulated that street names are an important means to communicate messages. Streets and street names are omnipresent in cities and, unlike statues, not restricted to a limited (elite?) part of urban space. The names can be placed in seven broad categories: names relating to the topographical site of the road (referring, for example, to an adjacent rivulet or orchard); neutral or poetic names (using names of birds, flowers or fruits); names relating to the social order or local community (for example, referring to a market place); names referring to the national context (such as the many national heroes); names referring to the local or regional context (such as local historical figures or characteristic buildings); names referring to international things; and names that merely indicate the function of the road (such as ring road) so that they can hardly be considered a symbol (Taal 2003: 156–9; cf. Nas 1993a: 28).

Street names reveal which interest group holds hegemonic power. 'The power of "nomination" […] is often the first step in taking possession [and street names are] among the first to undergo a refurbishing to commemorate new regimes' (Joel Richman quoted by Yeoh 1996: 221). Although Dutch names formed a minority – in 1940 only 21 per cent of the 1113 recorded street names in Surabaya were Dutch (Sarkawi 2004: 7) – most important streets bore a Dutch name. Although the Dutch language was forbidden

---

Medan to W.J.J.A. van de Bosch, vertegenwoordiger Bataafse Petroleum Maatschappij, Medan 28-11-1949, Arsip Medan, copyboek 5034/B.20.

during the Japanese period, street names were rarely altered. In Surabaya only one street (not even named after a Dutchman) was renamed. The Jalan Kenpeitai was named after the Japanese military police, which did not, however, have its headquarters there (Sarkawi 2004: 8).

According to both Nas (1993a: 29) and Taal (2003: 156), after independence streets in Jakarta and Palembang named after Dutchmen were given the names of their respective adversaries. For instance, the Van Heutszboulevard in Jakarta became Jalan Teuku Umar (Teuku Umarstreet) and the Julianalaan in Palembang became Jalan Kartini. I think that the hypothesis that Dutch names were changed for their opposites is not convincing and needs further study. Although Van Heutsz and Teuku Umar were the main protagonists in the Aceh–Dutch War, they never came face to face and partly belong to different phases of that war (Jonker 1917: 81–4). Kartini (an early twentieth-century role model of an emancipated woman) cannot reasonably be considered the opposite of Queen Juliana. I believe it is not enough to analyze the names themselves, and ideally one should know more about the motives for choosing a certain name. Fortunately, we are pretty well informed about toponymy in Medan.

Most of Taal's categories were encountered in colonial Medan (*Hind 1051 Medan* 1945; *Wegennet* n.d.). Some streets that referred to the site of the road were Paleisweg (passing the Sultan's palace), Kesawan (towards the wet ricefield), Bentengweg (Fortress Road), or Poloniaweg (leading to and through the former estate Polonia). Names derived from the flora fell into a neutral category. Marktstraat (Market Street) and Nieuwe Markt (New Market) refer to the social order or local community. A ward with streets named after Indonesian islands referred to the national context. Some major streets referred to the local context: Nienhuysweg and Cremerweg (named after two pioneers who laid the foundation for the plantation economy of Medan's hinterland), Sultansweg (Sultan's Road), Sultan Maämoen Alrasjidweg, and Burgemeester Mackaylaan (the first mayor of Medan). There were also two neighbourhoods named after members of the House of Orange and Dutch naval heroes. These last two examples show that Taal's categorization is not wholly adequate in a colonial situation, for do De Ruyterlaan, Evertsenlaan, Julianastraat, and Emmastraat refer to the national or international context? From the Dutch colonial perspective they are national names, but for all other ethnic groups at least the admirals must have been foreign.[23]

---

[23] The participation of all ethnic groups in royal holidays suggests that at least some of the members of the royalty were part of the national (pan-archipelago) consciousness.

The street names reveal that the municipal administration still had an explicit policy of ethnic segregation, which had to be reflected in toponyms.[24] Discussing some new streets, the administration remarked that a 'European' neighbourhood should bear European names, for example of local and regional (Dutch) administrators (*Gemeenteblad 2 Gemeente Medan* 1938: 146). In an indigenous ward (*kampung*), which looked like one large orchard, the streets were named after indigenous fruit trees (*Gemeenteblad 2 Gemeente Medan* 1931: 249) and here the word *djalan* (street) instead of the Dutch *straat* was used (*Hind 1051 Medan* 1945). In the Tamil ward streets were named after Indian cities (Calcutta, Madras, Bombay, Colombo) and the area with Chinese shopping streets bore Chinese names: Pekingstraat, Shanghaistraat, Kapiteinsweg and Luitenantsweg (the ranks of two formal Chinese leaders), and Hakkastraat (the name of one of the Chinese dialect groups).

Name changes reveal minor conflicts about the symbolism of the street names. One *gang* (alley) was named after the late *mantri* (police officer) Hadji Abdul Moetalib. His descendants objected that the word *gang* was belittling, therefore the name was changed to Mantrilaan (*Gemeenteblad 2 Gemeente Medan* 1930: 149). Chinese citizens, no doubt with a nationalist inclination, complained that the name Macaostraat was an insult and requested it be switched to Kongfoestraat. The request was granted after consultation with the *Majoor der Chinezen*, formal head of the Chinese community (*Gemeenteblad 2 Gemeente Medan* 1936: 99). The residents of the Baboerapark (Babura is the name of one of the two rivers in town) suggested their street be renamed Beatrixstraat, after the new-born Crown Princess. To lend their argument strength, they pointed out that confusion often reigned about the Baboerapark and Baboeraweg. The administration almost accepted the whole argument. Unfortunately for the royalist inhabitants of Baboerapark, the administrators thought that the Baboeraweg had more grandeur than the Baboerapark (which was a cul-de-sac) and was the one to be renamed Prinses Beatrixlaan (*Gemeenteblad 2 Gemeente Medan* 1938: 21).[25]

What can be concluded from these examples, apart from the fact that

---

[24] For Singapore, see Brenda Yeoh (1996: 224–8).

[25] The high regard for the House of Orange made urban administrators uncertain about the question whether royal names could be used freely. When the Manado administration wanted to baptize the bridge over the Tondano River – still under construction – Beatrixbrug, the mayor wondered whether permission from the royal family was required. Letter from Burgemeester Kapteijn to Adviseur voor Decentralisatie, Manado 18-2-1938, no. 5141, Arsip Nasional Republik Indonesia in Jakarta, Binnenlandsch Bestuur 1798.

many nags seemed to live in town? As was the case with statues and processions, the symbolic message was pluriform and there was room for input from below. Local participation was also possible when the administration named streets in an indigenous ward after consultation with the inhabitants (*Gemeenteblad 2 Gemeente Medan* 1934: 67), although sceptics may think that 'being consulted' will have meant no more than giving consent. The Sultan of Deli, who resided in Medan, also had a finger in the pie. In an elaborate agreement between municipality and Sultan about the use of Sultan's land for public housing, one of the articles stipulated that the Sultan retained the right to name the streets to be laid out in the area.[26]

At this point a caveat must be made, namely that the analysis of street names is based on official names. In my contemporary experience, Indonesians orient themselves more by marked points (mostly a specific building) than by a street name; the Chinese in Singapore also used landmarks for orientation more than the British colonials did (Yeoh 1996: 223, 231–2). At least we can expect that in colonial Indonesia many street names were translated into Malay. A map of about 1910 (*Wegennet* n.d.) uses these Malay names in a semi-official way: Marktstraat has become Djalan Pasar, Kapiteinsweg Djalan Kapitan and so on.

After the major shifts in power, the names were deemed inadequate in Medan. Street names referring to the House of Orange and administrators and institutes derived from Dutch colonialism were changed during the Japanese period. The Esplanade was baptized Lapang Foekoeraido (Loderichs 1997: 60–1). When the Dutch took over the city again in 1946, they restored the old names. Very soon after the transfer of sovereignty, the post-independence administration of Medan felt in its turn that many street names were no longer appropriate to the new situation. Six indigenous people (five of whom were civil servants) and one Chinese person (member of the provisional city council) were invited to sit on a committee to consider the names. The mayor had names of regions in Medan's hinterland (that is: neutral names or names from the regional context) in mind for an extension of the city.[27] Already by mid-1950, official correspondence used an Indonesian name, with the Dutch colonial name between brackets.

---

[26] Afschrift overeenkomst tussen Sultan van Deli en Rijksgroten en gemeente Medan, 3-4-1930, attached to letter gemeente to Negara Soematera Timoer, Medan 31-5-1948, Arsip Medan, copyboek 2207.

[27] Surat walikota Medan 26-1-1950, Arsip Medan, copyboek no. 439/DC-2; Notulen adviesraad stadsgemeente Medan, 29-11-1949, attached to brief gemeentesecretaris aan raadsleden 29-12-1949, Arsip Medan copyboek 5544/A-12.

A list of 105 old and new names (Djawatan Penerangan Medan 1954: 216–18) shows that neutral names were not the only designations chosen; names from the Revolution also appeared. However, streets named after Dutch soldiers were not changed to their direct adversaries, although some Dutchmen were replaced by Indonesians from a different time but the same region (Van Heutsz-Iskandar Muda; Bontekoe-Imam Bondjol; Coen-Gadjah Mada; Daendels-Hajam Wuruk).[28] The elite neighbourhood with naval heroes had their names replaced by those from the Independence struggle: Djalan Gerilla, Djalan Renville, Djalan Linggardjati, Djalan Ratulangi, Djalan Sudirman, Djalan Tjut Nja Din, Djalan Chairil Anwar.[29] A small ward with European scientists was given the names of heroes from Malay epics (so Melchior Treublaan became Djalan Hang Tuah). Orange names adjacent to the Chinese ward were given Chinese names – Djalan Taipeh, Djalan Sun Yat Sen – so that the Chinese area was extended. Where no coherence had existed in the past, it was introduced; in a ward with motley names streets were now consistently named after temples (Djalan Tjandi Borobudur, Djalan Tjandi Prambanan, et cetera). Finally, a number of street names were translated in a freely associating style: Marktstraat became Djalan Perdagangan, Nieuwe Markt became Djalan Perniagaan, Burgemeester Mackaylaan became Djalan Walikota, and Max Havelaarlaan became Djalan Multatuli. Of the main roads, only the Moskeestraat was literally translated (to Djalan Mesjid). Streets that already had an Indonesian name (of a fruit or island for example) were not changed.

Most important of all changes in Medan was that street names referring to the Dutch colonial presence were removed, but not altered to their opposites. The new names did not represent a confrontation with colonialism, but attempted gently to write the Dutch out of the urban landscape. It seems that this rather detailed analysis of street names in Medan reflects a general pattern found in other cities as well.[30] Dutch names

---

[28] Captain Bontekoe blew up his ship off the west Sumatran coast; the other names are obvious to people with a basic knowledge of Indonesian history. I do not know whether it was policy to pick new names from those Indonesian people from the same region, but a different era.

[29] *Gerilla* is Indonesian for guerrilla warfare; Renville and Linggardjati are places were two peace agreements were signed, Ratulangi was a moderate nationalist, Sudirman was the by then already late commander of the Revolutionary army, Tjut Nja Din was a nineteenth-century female resistance leader, and Chairil Anwar was a poet of resistance songs.

[30] This generalization applies to Surabaya (*Gewijzigde Straatnamen* n.d.), Bandung (*Perubahan Nama Djalan-djalan* 1950) and Semarang (*Nieuwe Straatnamen* n.d.). In Bandung, a few European names considered all right survived the name change: Curie, Van Deventer, Professor Eykman, Pasteur; the name Multatuli was moved to a smaller street.

remained, however, for a long time in advertisements. At least until 1955 signboards with Dutch words, Chinese characters, and even Dutch flags were still visible in the street (see for example the photos in Broeshart *et al.* 1994: 91, 112; Voskuil *et al.* 1996: 124, 136–43).

Indonesian symbolic nationalism was gradually extended to other non-indigenous minorities. In 1957 the administration of Medan formed a commission to reconsider the street names in the Chinese neighbourhood.[31] Nowadays, not only the Chinese names have disappeared, the Indian ones (like Calcuttastraat) suffered the same fate. Sarkawi (2004: 12–13) has demonstrated that names in the Chinese neighbourhood in Surabaya began to be changed to Jalan Coklat, Jalan Kopi (Chocolate Street, Coffee Street) and others in 1958. The name change was settled by the municipal council in 1962. So, the anti-Chinese sentiment was established in different places at about the same time.[32] Less than a decade earlier the portrait of Mao Zedong had still been carried along in parades on Labour Day (van Schaik 1996: 86).

It is interesting that Chinese names were removed at about the same time that the remaining Dutch statues were demolished. I do not know why, but it is possible that a reinvigorated nationalism was promoted by politicians who wished to distract people's attention from domestic problems (the political stalemate after the 1955 election), growing irritation with the unsolved West Irian issue, and the continued grip of foreign capital on economic affairs (Cribb and Brown 1995; Feith 1962). It is a matter of perspective whether the disappearance of Chinese and Indian symbols is interpreted as suppression of an ethnic identity or as the gradual loss of the meaning of ethnicity. The former interpretation is the most likely; in 1965, anti-Chinese feelings were virulent.

At the end of this section it should be remembered that we know little about unofficial street names. Sarkawi (2004: 2–6) gives a rare glimpse of

---

These names remained in use even after the Indonesian–Dutch relationship deteriorated over the status of West New Guinea in the early 1960s (*Nama Djalan dan Peta Bandung* 1966).

[31] Letter M. Tahir, Wakil Sekretaris Medan to Ketua Dewan Pemerintah Daerah Peralihan Propinsi Sumatera Utara 11-6-1957, Arsip Medan, copyboek 7239/DC-2.

[32] Chinese names still figured in lists of streets in Bandung and Surabaya of about 1966 (*Nama Djalan dan Peta Bandung* 1966; *Nama Djalan dan Peta Surabaja* 1966). The discrepancy with Sarkawi's convincing analysis may be explained by the fact that the lists of street names, which do not bear a year, have been assigned a wrong year of publication by the KITLV librarian. In 1958, the Public Works Department had proposed to replace the Chinese names in Surabaya by the names of the *wali songo*, the nine men who introduced Islam in Java, but this was apparently too much of an affront to the Chinese community, so that eventually the neutral names of chocolate and tea were chosen.

how public opinion in Surabaya differed from the official names. The urban administration promulgated in both 1918 and 1955 by-laws claiming the monopoly to name streets, but the administrative hegemony did not stop ordinary people using the popular names. For example, in 1961 the Surabaya administrators changed the name Jalan Raya Darmo, the main road in town, to Jalan Patrice Lumumba, one month after the popular Congolese leader was murdered. The administrators chose to change the name rather to please President Sukarno than out of conviction. The same year, however, the government restored the old name, because of the many popular protests against the change of name; the name Jalan Patrice Lumumba was given to a section of a smaller street. In another example, the 17 August Commission asked the government to change Jalan Pemuda (Young Revolutionaries Road) into Jalan Pahlawan (Heroes Road), and to change the name of the already extant Jalan Pahlawan into Jalan Merdeka (Freedom Road). The city administration fulfilled the first proposal, but instead of satisfying the second proposal too, it decided to change the extant Jalan Pahlawan back to the old name, Jalan Simpang.[33] The old name was still commonly used, also in compounds, such as Simpang Hospital. Sarkawi concludes that people usually wish to preserve the old names, whereas the government introduces new toponyms and wants the people to forget the old names. In 1968, the Surabaya government even issued a circular mentioning the names that no longer conformed to the formal names and were therefore forbidden. Of course, such a circular could not be enforced. Whether the disregard for official names rendered the ordinary people less open to the surveillance of the (colonial) state, as Brenda Yeoh (1996: 234) contends, is, I believe, an unproven assumption.

Conclusion

At the beginning of this article I argued that the study of urban symbolism should become more theoretical by being put on a comparative footing, a return to an emic approach, and a focus on processes of symbol creation. At the end of the day, it has appeared easier to criticize others than to write something constructive myself. Comparison of Indonesian cities has resulted in empirical generalizations, but not in new theoretical insights. For example, there is still no clear explanation of why Dutch statues were demolished in Padang but spared in Medan. The participants' viewpoint has been obscured by the historical distance to the subject of study, although some sources provided evidence of participants' ideas; the arguments for a

---

[33] Nowadays, Jalan Simpang is officially called Jalan Pemuda, but the old name is still used.

change of street name are a good example. The historical analysis has only partially met the plea for a focus on processes of symbol creation.

The main question of the historical analysis was how political watersheds in Indonesia (Japanese occupation and independence) made an impact on the creation of and change in urban symbols. Notwithstanding the colonial power, the message sent by statues in Dutch times was surprisingly pluriform. The Dutch were dominant, but not hegemonic.[34] The pluriformity was formed by the fact that many statues were created, and funded, on the initiative of common people and not by the state. This pluriformity survived the upheavals of war and revolution to a considerable degree: Dutch statues were not always directly destroyed and Japanese and Indonesian people added new colours to the symbolic palette.[35]

Street names, in contrast, were altered quickly. Changes were aimed at the removal of Dutch symbolic presence rather than creating a nationalist image. Because only the state gave street names, the symbolism was more hegemonic than was the case of statues. But the hegemonic removal of Dutch names was undermined in several ways. Dutch names remained common on signboards for many years. Folk names of streets could exist alongside official names. Old official names remained in use long after they have been formally abolished.

Compared to the pluriform and resilient message of statues, and the quickly implemented but undermined change of street names, the symbolic changes in parades and flags were quick and total. Parades drew much attention, more perhaps than the statues, because they were a break in the daily drudgery. The parades, mobilizing people, and the act of putting out a flag required the active participation of the audience and could not be ignored like street names. The changes in parades and flags were also more abrupt than the other types of symbol carriers discussed here. This abruptness is, I believe, the reason why hoisting a flag can be very emotional and is well remembered. The renowned flag incident in Surabaya (September 1945) has even become a symbol of the much bloodier, but unrelated Republican–British urban war of November 1945. Finally, parades and flags cost less than a statue. A flag is a poor person's monument.

---

[34] This pluralism was also visible in the important religious processions, associated with particular ethnic groups, which have not been discussed in any detail here. They drew large crowds, including those of other religions, which underscored a certain mutual tolerance and recognition.

[35] The plethora of nationalist statues stems from the 1960s (in Jakarta) and 1980s and 1990s (in provincial capitals) and calls for a different explanation, which falls outside the scope of this article.

Symbolic changes were not only discernible for individual types of symbol carriers, but a shift also occurred between the types. This brings us to the last question to be answered, namely which of the material, behavioural or discursive signifiers were most powerful. The reasons for the shift from statues to parades go further than the practicalities of time and funds available to erect a statue. An additional explanation is found in the differential basis and perception of power by Europeans and Indonesians.[36] Material signifiers were most important in Dutch colonial times, which was in accordance with a European culture of petrified images (not only of statues but also of imposing offices not discussed here) and a military strategy aimed at the occupation of territory. The required deference towards the Japanese overlords could not be shown in statues, which were not part of Japanese culture anyway, but had to be expressed by behavioural signifiers. Military parades were a show of arms on the part of the Japanese and the ordinary people responded by respectful behaviour consisting of bows and a show of flags. In the Malay tradition, power was also not expressed in statues but in the number of followers. Parades were important, not as a show of arms, but because of the large rallies of spectators. The power to confer new names (on followers in pre-colonial times or streets in the 1950s) was more important than the new names as such.

In short, the political turns were reflected in changes in different types of symbol carriers. The relative shift between the types, from statues to parades, is connected to the different meaning that Dutchmen, Japanese and Indonesians ascribed to those types of symbol carriers. The decline of the statue and the rise of the parade was a sign of the times.

References

Anderson, B.R.O'G. (1992) *Long-distance Nationalism: World Capitalism and the Rise of Identity Politics*. Amsterdam: Centre for Asian Studies Amsterdam.

Bain, L. (2002) '"Indonesia", dari sebuah hotel.' *Kalam*, vol. 19, pp. 35–58.

Broeshart, A.C., J.R. van Diessen, R.G. Gill and J.P. Zeydner (1994) *Soerabaja: Beeld van een Stad*. Purmerend: Asia Maior.

Budiman, A. (1979) *Semarang Juwita: Semarang Tempo Doeloe Semarang Masa Kini dalam Rekaman Kamera*. Semarang: Tanjung Sari. [Jilid 1.]

Colombijn, F. (1994) *Patches of Padang: The History of an Indonesian Town in the Twentieth Century and the Use of Urban Space*. Leiden: Research School CNWS.

——— (1998) 'Canberra: A sheep in wolf's clothing.' In: P.J.M. Nas (ed.), Special issue: Urban rituals and symbols. *International Journal of Urban and Regional Research*, vol. 22, pp. 565–581.

---

[36] I cannot go into this topic here, but the concept of the theatre state (Geertz 1980) and the importance of regalia (Gullick 1988: 45–6) could be a starting point for further analysis.

—— (2001) 'East Timor, from ashes to nationhood: Interview with José Ramos-Horta.' *IIAS Newsletter*, vol. 25, pp. 4–5.

Cribb, R. and C. Brown (1995) *Modern Indonesia: A History since 1945*. London: Longman.
Deli Courant.

Dick, H.W. (2002) *Surabaya, City of Work: A Socioeconomic History, 1900–2000*. Athens, Ohio: Ohio University Press.

Dinas Museum dan Pemugaran (1999/2000) *Monumen dan Patung di Jakarta*. Jakarta: Dinas Museum dan Pemugaran Propinsi DKI Jakarta.

Dinas Museum dan Sejarah (1980) *Sejarah Singkat Pendirian Patung-patung dan Monumen di Jakarta*. Jakarta: Dinas Museum dan Sejarah, Pemerintah DKI Jakarta.

Djawatan Penerangan Medan (1954) *Buku Tahunan Kota Besar Medan Tahun 1954*. Medan: Djawatan Penerangan Kota Besar Medan.

Dorrestein, R. (2000) *Het Geheim van de Schrijver*. Amsterdam and Antwerpen: Contact.

Evans, G. (1998) *The Politics of Ritual and Remembrance: Laos since 1975*. Honolulu: University of Hawai'i Press.

Feith, H. (1962) *The Decline of Constitutional Democracy in Indonesia*. Ithaca: Cornell University Press.

Firth, R. (1973) *Symbols, Public and Private*. London: Allen and Unwin.

Frederick, W.H. (1999) 'Weerspiegeling in stromend water: Indonesische herinneringen aan de oorlog en de Japanners.' In: R. Raben (ed.), *Beelden van de Japanse Bezetting van Indonesië: Persoonlijke Getuigenissen en Publieke Beeldvorming in Indonesië, Japan en Nederland*, pp. 16–35. Zwolle: Waanders and Amsterdam: Nederlands Instituut voor Oorlogsdocumentatie.

Geertz, C. (1980) *Negara: The Theatre State in Nineteenth-century Bali*. Princeton: Princeton University Press.

Gemeenteblad 2 Gemeente Medan.

Gewijzigde Straatnamen (n.d.) *Gewijzigde Straatnamen van de Stad Soerabaia*. Sourabaia: Nieuwe Courant.

Glaser, B.G. and A.L. Strauss (1967) *The Discovery of Grounded Theory: Strategies for Qualitative Research*. Chicago: Aldine.

Gullick, J.M. (1988) *Indigenous Political Systems of Western Malaya*. London and Atlantic Highlands, N.J.: Athlone Press. [Revised edition.]

Han Bing Siong (2003) 'Captain Huyer and the massive Japanese arms transfer in East Java in October 1945.' *Bijdragen tot de Taal-, Land- en Volkenkunde*, vol. 159, pp. 291–350.

Heine-Geldern, R. (1942) 'Conceptions of state and kingship in Southeast Asia.' *Far Eastern Quarterly*, vol. 2, pp. 15–30.

Hind 1051 Medan (1945) *Hind 1051 sheet Medan*, 1: 5,000. Washington: War Office. [Second edition.]

Indurthy, R. (1994) 'The demolition of the Babra Mosque in Ayodhya and the implications for India's secularism and political stability.' *Asian Profile*, vol. 22, no. 1, pp. 51–65.

Jeffrey, R. (1980) 'What the statues tell: the politics of choosing symbols in Trivandrum.' *Pacific Affairs*, vol. 53, pp. 484–502.

Jezernik, B. (ed.) (1999) *Urban Symbolism and Rituals: Proceedings of the International Symposium Organised by the IUAES Commission on Urban Anthropology, Ljbuljana, June 23–25, 1997*. Ljubljana: Univerza v Ljubljani.

Jonker, C.G.J. (1917) 'Atjèh.' In: J. Paulus (ed.), *Encyclopaedie van Nederlandsch-Indië*, eerste deel, pp. 68–89. 's-Gravenhage: Martinus Nijhoff and Leiden: E.J.Brill. [Tweede druk.]

Lapian, A.B. (1999) 'Onvoltooid monument.' In: R. Raben (ed.), *Beelden van de Japanse Bezetting van Indonesië: Persoonlijke Getuigenissen en Publieke Beeldvorming in Indonesië, Japan en Nederland*, pp. 105–6. Zwolle: Waanders and Amsterdam: Nederlands Instituut voor Oorlogsdocumentatie.

Leclerc, J. (1993) 'Mirrors and the lighthouse: A search for meaning in the monuments and great works of Sukarno's Jakarta, 1960–1966.' In: P.J.M. Nas (ed.), *Urban Symbolism*, pp. 38–58. Leiden, New York and London: E.J. Brill.

Lee Kuan Yew (2000) *From Third World to First: The Singapore Story: 1965–2000*. New York: HarperCollins.

Loderichs, M.A., with contributions by D.A. Buiskool, B.B. Hering *et al.* (1997) *Medan: Beeld van een Stad*. Purmerend: Asia Maior.

Madjalah Kota Medan.

Maier, H. (1999) 'De wind zal blazen: Moderne Indonesische literatuur over de Japanse tijd.' In: R. Raben (ed.), *Beelden van de Japanse Bezetting van Indonesië: Persoonlijke Getuigenissen en Publieke Beeldvorming in Indonesië, Japan en Nederland*, pp. 89–104. Zwolle: Waanders and Amsterdam: Nederlands Instituut voor Oorlogsdocumentatie.

Nama djalan dan peta Bandung (1966) *Nama Djalan dan Peta Kota Raya Djakarta, Kota Madya Bandung*. Djember: Sumber Ilmu.

Nama djalan dan peta Surabaja (1966) *Nama Djalan dan Peta Kota Madya Surabaja, Kota Madya Malang*. Djember: Sumber Ilmu.

Nas, P.J.M. (1990) 'Jakarta, stad vol symbolen: Met Leiden als contrast.' *Antropologische Verkenningen*, vol. 9, no. 3, pp. 65–82.

—— (1993a) 'Jakarta, city full of symbols: An essay in symbolic ecology.' In: P.J.M. Nas (ed.), *Urban Symbolism*, pp. 13–37. Leiden: E.J. Brill.

—— (ed.) (1993b) *Urban symbolism*. Leiden: E.J. Brill.

—— (ed.) (1998a) Special issue: Urban rituals and symbols. *International Journal of Urban and Regional Research*, vol. 22, pp. 545–622.

—— (1998b) 'Introduction: Congealed time, compressed place: roots and branches of urban symbolic ecology.' In: Special issue: Urban rituals and symbols. *International Journal of Urban and Regional Research*, vol. 22, pp. 545–49.

—— (2003) 'Ethnic identity in urban architecture: Generations of architects in Banda Aceh.' In: R. Schefold, P.J.M. Nas and G. Domenig (eds), *Indonesian Houses: Tradition and Transformation in Vernacular Architecture*. Leiden: KITLV Press. [Volume 1.]

Nas, P.J.M. and M. van Bakel (1999) 'Small town symbolism: The meaning of the built environment in Bukittinggi and Payakumbuh (West Sumatra, Indonesia).' In: B Jezernik (ed.), *Urban Symbolism and Rituals: Proceedings of the International Symposium Organised by the IUAES Commission on Urban Anthropology, Ljbuljana, June 23-25, 1997*, pp. 173–89. Ljubljana: Univerza v Ljubljani.

Nas, P.J.M. and R. Sluis (2002) 'In search of meaning: Urban orientation principles in Indonesia.' In: P.J.M. Nas (ed.), *The Indonesian Town Revisited*, pp. 130–46. Münster: Lit and Singapore: Institute of Southeast Asian Studies.

Nelson, J. (2003) 'Social memory as ritual practice: Commemorating spirits of the military dead at Yasukuni Shinto Shrine.' *The Journal of Asian Studies*, vol. 62, pp. 443–67.

Nieuwe Straatnamen (n.d.) *Nieuw Straatnamen Semarang en Salatiga: Daftar Nama-nama Djalan Baru di Kota Semarang dan Salatiga*. Semarang: De Locomotief.

Panikkar, K.N. (1993) 'Religious symbols and political mobilization: the agitation for a

Mandir at Ayodhya.' *Social Scientist*, vol. 21, no.7–8, pp. 63–77.

Pelaksana Pembina Tugu Nasional (1978) *Tugu Nasional: Laporan Pembangunan*. Jakarta: Pelaksana Pembina Tugu Nasional.

Perubahan nama djalan-djalan (1950) *Perubahan Nama Djalan-djalan di Bandung: Gewijzigde Straatnamen van Bandung, 1950*. Bandung: De Preangerbode.

Phillips, L.W. (ed.) (1984) *Ernest Hemingway on Writing*. New York: Charles Scribner's Sons.

Pieke, F. (1993) 'Images of protest and the use of urban space in the 1989 Chinese people's movement.' In: P.J.M. Nas (ed.), *Urban Symbolism*, pp. 153–71. Leiden: E.J. Brill.

Raben, R. (1999) 'Overlevering in drievoud.' In: R. Raben (ed.), *Beelden van de Japanse Bezetting van Indonesië: Persoonlijke Getuigenissen en Publieke Beeldvorming in Indonesië, Japan en Nederland*, pp. 7–14. Zwolle: Waanders and Amsterdam: Nederlands Instituut voor Oorlogsdocumentatie.

Sarkawi B.H. (2004) 'Sepanjang jalan kenangan: Makna dan perebutan simbol nama jalan di kota Surabaya.' Paper presented at the First International Conference on Urban History in Indonesia, Surabaya, 23–25 August.

Schaik, A. van (1996) *Malang: Beeld van een Stad*. Purmerend: Asia Maior.

Scott, J.C. (1985) *Weapons of the Weak: Everyday Forms of Peasant Resistance*. New Haven: Yale University Press.

Soerabaiasch-Handelsblad.

Stutterheim, W.F. (1948) *De Kraton van Majapahit*. 's-Gravenhage: Martinus Nijhoff.

Suti, B. (1979) *Medan Menuju Kota Metropolitan*. Medan: Yayasan Potensi Pembangunan Daerah.

Taal, S. (2002) 'Cultural expressions, collective memory and the urban landscape in Palembang.' In: P.J.M. Nas (ed.), *The Indonesian Town Revisited*, pp. 172–200. Münster: Lit Verlag and Singapore: Institute of Southeast Asian Studies.

—— (2003) *Between Ideal and Reality: Images of Palembang*. Leiden: Leiden University. [PhD thesis.]

Visser, J.G. (1979), 'Groot, Cornelius Johannes de.' In: J. Charité (ed.), *Biografisch Woordenboek van Nederland*, Eerste deel, pp. 214–17. 's-Gravenhage: Martinus Nijhoff.

Voskuil, R.P.G.A. (1993) *Batavia: Beeld van een Stad*. Purmerend: Asia Maior. [Tweede herziene uitgave.]

—— with contributions by C.A. Heshusius, K.A. van der Hucht, V.F.L. Pollé and H.G. Spanjaard (1996) *Bandoeng: Beeld van een Stad*. Purmerend: Asia Maior.

Wegennet (n.d.) *Wegennet der Gemeente Medan*, 1:10.000. [Unknown publisher.]

Yeoh, B.S.A. (1991) 'The control of "sacred space": Conflicts over the Chinese burial grounds in colonial Singapore, 1880-1930.' *Journal of Southeast Asian Studies*, vol. 22, pp. 282–311.

—— (1996) *Contesting Space: Power Relations and the Urban Built Environment in Colonial Singapore*. Kuala Lumpur: Oxford University Press.

Chapter 7

UNDER THE TABLECLOTH
Exploring symbolism in central Cape Town

Peter J.M. Nas, Margot N. te Velde, Annemarie Samuels

Introduction[1]

Cape Town is located on the Cape Peninsula, the southernmost tip of South Africa. Mountains and the sea enclose it. Nobody knows exactly when the first people came to this place, but the first time European explorers set ashore at the Cape, they encountered an indigenous people, the Khoisan. The Portuguese were the first Europeans who discovered the Cape, in their search for a route to India, but they did not stay there. The Dutch, on the contrary, settled in Cape Town, which remained under Dutch rule for a century and a half.

In 1652, the Dutch commander, Jan van Riebeeck, established a post of the East India Company at the Cape, to provision ships with fresh food. The settlement became a small town and because the Dutch thought the indigenous Khoisan people unsuitable for labour, they imported slaves from Southeast Asia. Their descendants are now known as Cape Malays, a

---

[1] This article is an abridged version of the unpublished MA thesis *Between Devil's Peak and Lion's Head: Urban Symbolism in Cape Town, South Africa* written by Margot N. te Velde in 1998 (Leiden University) under supervision of Prof Dr Peter J.M. Nas. This thesis is based on the theory about urban symbolism developed by Peter Nas and fieldwork in central Cape Town by Margot te Velde from October 1997 to February 1998. The peripheries of Cape Town could not be covered in this study, because of unfavorable fieldwork conditions. As Margot te Velde did not have the opportunity to rework her thesis for publication into an article, she kindly agreed that this task could be carried out by Peter Nas and Annemarie Samuels under reference of the three persons involved. In order to include the most recent developments, a section containing information on the last ten years was added by Margot te Velde in 2004.

Muslim community. In 1806, a British force ended Dutch rule after a short battle and took over Cape Town. In the following century the city expanded, and industry and commerce developed. The Britons introduced their culture to the city and a strict social hierarchy was established in which ethnicity played an important role. During the twentieth century the ideology of apartheid transformed Cape Town. People who had been part of the city for centuries and whose birthright it was to be there were forcibly moved to the Cape Flats at the outskirts of the city.

After the National Party election in 1948, total separation of black and white at every level became official policy, but its foundations had already been laid earlier by the segregation policy of the British Government at the beginning of the twentieth century. Cape Town was planned as an apartheid city. Black, coloured and white areas where divided by railway lines and highways. This segregation did not disappear after the ending of apartheid. Today it could be said that several Cape Towns exist side by side, separate and unequal, in which people are divided by ethnicity, economic position, language, party politics, culture and religion.

In this study of Cape Town symbolism, we will explore present-day, human-made urban symbols, such as statues, street names and architecture, but we are also looking at the natural environment and its landmarks, which prove to be of special importance. A landmark in this context is an element in the urban landscape that a resident sees regularly, often on a daily basis, which makes a place familiar and unique, and is a possible point of reference. In our description of Cape Town's symbolism we will stress these two levels, namely natural environment and history. The Devil's Peak, Table Mountain, Lion's Head and Table Bay make up the former. The latter is divided in four periods: the Dutch colonial time (1652–1806), British rule (1806–1918), the years of apartheid (1918–1994) and the period of transition and modernization (1994 to the present).

Natural environment

The natural environment of Cape Town, especially the mountain chain, plays an important role in the way Capetonians perceive their city. The Cape Peninsula is dominated by the Table Mountain massif with two attendant peaks called Lion's Head and Devil's Peak. The mountain chain that stretches over the entire Cape Peninsula forms a 1,100-metre-high wall that embraces Cape Town and separates the city from the inland (Figure 1).

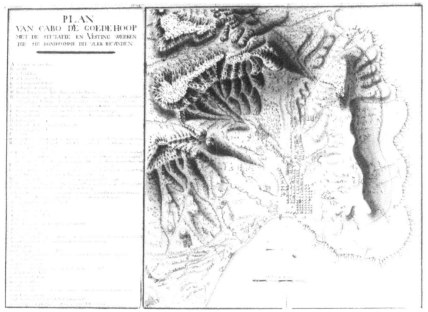

Figure 1 Map of Cape Town (Source: The Hague States Archives, VEL/839).

The mountains strongly influence the weather in Cape Town. In order to forecast their weather, Capetonians look at them, especially at the pyramid-shaped Lion's Head. When a ring of clouds hangs around its top, popularly called the 'Mountains Hat', rain is predicted for the following day. In summertime, clouds slide down the mountain disappearing when they approach the city. This is called the 'Tablecloth'. An urban myth which is common among Capetonians tells us about the origin of this phenomenon.[2] It relates how the retired pirate and old smoker Van Hunks came to the mountains, which he climbed when the weather was good. In summer he would light his pipe in the mountain and send clouds of blue smoke up from his pipe. One day a stranger sat himself down next to Van Hunks. He said how good it was to sit there and smoke, addressing Van Hunks by name. Van Hunks offered him a smoke and the stranger challenged him to smoke more than he could in a single sitting. Van Hunks took up the challenge with his soul as a stake against a fortune in red gold. They started smoking and gradually a choking white cloud grew around them. The next morning the top of Table Mountain was hidden in clouds which slid down into the city. The stranger asked Van Hunks for a drink. As he took a sip, his cap fell off

---

[2] For an extended version of the legend see Miller (1979).

his head and two sharp horns became visible. Van Hunks recognized him as the Devil. After that a flesh of lightning appeared and when the fog cleared the stranger and Van Hunks were gone. From that day on the area has been known as Devil's Peak. When the Southeast wind blows and a white cloud, called the Tablecloth, appears, Capetonians say that the Devil and Van Hunks are on the mountain again.

For Capetonians, Table Mountain is an ever-present reality. It dominates life in the city and means many things to many people, but is invariably seen as part of the city's identity. The mountain is considered a natural monument and something that distinguishes Cape Town from other cities. Table Mountain is the first thing people mention when asked to describe Cape Town in a physical sense. They see it as a familiar landmark as they approach the city. It is a landmark indicating they are almost home.

The mountain is related to all social, ethnic, political, cultural and religious groups and in so being it is an encompassing, neutral symbolic feature belonging to all the inhabitants of Cape Town. It is seen as a natural monument, a point of natural beauty, but also as an indefinite wilderness. Homeless people are called *Bergies*, which means people of the mountain, and are said to have lived in the wilderness of the mountains before entering town. With more than 1,400 indigenous species of plants, Table Mountain is awaiting designation as a world heritage site and abounds in recreational and eco-tourist value.

Table Mountain has also been a monument to continuity, while the city has seen much change through the centuries. But, it is dynamic and controversial as well. Nowadays, there are many groups that campaign against the development of the mountain. They want to save the environment from being spoiled by tourism and modernization, and they want to keep Table Mountain intact as a heritage for their children, as a specific part of Cape Town's identity.

In addition, the natural environment of Cape Town also has a functional value. We recognize the shape of Table Mountain, Devil's Peak and Lion's Head in the logo of Cape Town. This logo is symbolically promoting unity and uses the peaks as the unique mark of the city to demonstrate its identity to the rest of the world (Figure 2).

Travel guides present an image of Cape Town as a place of great natural beauty blessed by mountains, sun, sea and beaches. They reveal Cape Town as a natural wonder. Capetonians seem to experience their city in this way as well. Of course, crime and housing are mentioned as important social problems, but in their description of the city Capetonians refer primarily to natural beauty. They mention the sea, the mountain and the beaches.

Pertinently, they seem to portray their city as a place of exclusivity. The text of a popular car sticker says: 'Welcome to Cape Town – now go home'. It is a warning to the people from Transvaal in the north of South Africa who spend their summer holidays at the Cape not to outstay their welcome.

Figure 2 Table Mountain (Photograph: Margot te Velde, 2004).

Indisputably, for Capetonians natural beauty seems to be a very important aspect of their city. The natural environment is imbued with various meanings, and is related to different values and functions. It clearly constitutes an important part of Cape Town's identity.

The Dutch colonial period

The Dutch used the Cape of Good Hope as a provisioning port for the ships of the Dutch East India Company on their voyage to the East Indies and back home. The site was considered a strategic holding rather than a centre of trade, production or administration. The planning of the settlement is visible in the grid plan, which was designed in Amsterdam and established in Cape Town at the end of the seventeenth century.

Historically, in general the grid has served two main purposes: to facilitate orderly settlement and to promote modernization in contrast to what was considered to be chaos (Kostof 1991: 102). Geometrical forms dominated the layout of Cape Town in the Dutch period. The town was a rectangle, located between the fort, the Company's Garden and Lion's Head. The streets intersected rectangular blocks. Low-walled canals that resembled those in Holland completed the grid. The similarity with Dutch towns caused Cape Town to be referred to as 'Little Amsterdam'. But behind the proud town houses, less readily controlled by the orderly geometrical landscape, were the side streets and alleyways housing a large underclass of poor whites, artisans, free blacks, and slaves.

Considering Southall's (1993: 380) question about whether the rectangular city can be seen as 'a symbolic statement of human culture triumphing over nature', the case of Cape Town seems to suggest a positive answer. In addition, the Dutch grid cannot only be perceived as a symbol of domination over the natural environment, but also as a symbol of authority over the 'uncivilized' indigenous people.

The Dutch built a fort, the Castle of Good Hope, which is now the oldest surviving building in South Africa. First a square, it was replaced by a pentagonal construction in 1666. The fort was used as the Governor's residence, as a command post for observation of Table Bay and the town, and it housed a church and military offices. There are a dungeon and garrison cells (for the slaves) that were used to detain prisoners from passing ships on their way to the East during the Second World War. The main gate, built in 1682, resembles the gateway of the town of Dordrecht in Holland and the bastions of the Castle have Dutch names.

Nowadays, tourists can observe the changing of the guards at the Castle, which is preceded by the firing of the cannon on Signal Hill. The Castle is a symbol of dependency, of Dutch military and political control. It is used for exhibitions, as a military museum and an art gallery. But it has acquired new connotations as well. The dungeon now functions as a memorial site to the victims of apartheid and in this way the monument has become part of the country's spirit of unity and reconciliation.

The Company's Garden is a park, associated with the founding of the city. It was established by Jan van Riebeeck to supply passing ships with fresh vegetables and resembled contemporary Dutch gardens in the way it was divided into beds. The design was symmetrical and geometrical and the beds were based on the shapes of the rectangle and the square. The Cape garden resembled the princely garden at the Dutch Huis ten Bosch in 1656 in the way the kitchen gardens were organized, but it was devoid of ornaments such as fountains, statues and parterres. This emphasized functionality and utility of the Cape Garden. However, the Britons erected statues that will be discussed later.

The location of the Garden was carefully planned. Table Mountain forms the end of a long axis on which the Garden was laid out and it actually feels as if the mountain is part of the garden. On the other side, the axis leads to the bay. The Garden is enclosed by Devil's Peak and Lion's Head (Fagan 1985: 134). The central axis of the garden formed the main axis of the early city and it is still one the most important streets in Cape Town.

Assailed by bad weather conditions that destroyed the crops, the Garden soon lost its function as a vegetable garden and was turned over to recreation and botanical research. By the late eighteenth century, institutions such as a hospital, a church and a slave lodge were built on the edges of the Garden and, from 1860, more public institutions such as the South African Library and the Houses of Parliament were established in the vicinity.

The right of access to the Garden has changed throughout the centuries and the use of this public space is telling. The Dutch established the Garden and opened it to the public. Following the institution of British rule in 1806, access changed. The British Governor used the Garden's Tuynhuys as a private residence and closed almost half of the Garden to public use. Following the outcry by angry citizens, it was soon re-opened, but in 1950 access was denied to black and coloured people. Nowadays, the Garden is open to the general public again and used as a recreational place. Its major drawback is its being considered unsafe, especially after dark. A survey of the City Planner's Department in 1992 showed that people hold the Garden in high esteem because of the link with the past and its function as a meeting place. They view the Garden as a landmark of Cape Town.

Street names in the central part of Cape Town mark continuity, as they have undergone little change over the centuries. They especially reflect the Dutch and the British periods. During the time of van Riebeeck the settlement just had three streets and in 1723 Francois Valentyn mentioned that there were eight streets. In Dutch eyes, Cape Town was primarily a functional point for provisioning and they did little to develop the town. Streets were officially named only five years before the passing of the town to British rule. The Dutch primarily used descriptive names, such as Buitengracht, Waterkant and Kerkstraat. For entire areas the names of persons were used, such as Riebeeck's Kasteel and Simonsbaai.

As the presence of a new regime often results in radical changes in street names, it is surprising that the names of central Cape Town generally remained the same. The one change was that they were translated; for example, Plijnstraat became Square Street. Today, many street names are British translations of the original Dutch names. There are several names that refer to persons who played an important role in the city's past, such as St George's Street and van Riebeeck Square. These names are important elements in the 'collective memory'.

Almost all other street names in the city centre are related to the natural environment, such as Long Street and Greenmarket, and therefore belong to the 'cultural memory'. There are no Africanized street names. The names are English or Dutch. Districts in the centre have Dutch names relating to the environment and further from the centre districts have predominantly English names relating to the environment. Only in the Cape Flats are African names of streets and districts found.

There are no street names that relate to the 'historical memory', nor have street names changed since the abolition of apartheid. So, there are no names (yet) of anti-apartheid leaders. The street names in central Cape Town keep the memory of the Dutch and British rule.

The architecture found in the city is full of contrasts. Contemporary and old Dutch colonial architecture are mixed. Some of the modern buildings in Cape Town are controversial, such as the three cylindrical residential towers at the foot of Devil's Peak erected in 1968. They dominate the landscape and some Capetonians think they are a visual degradation of the place. Popularly, they are called the 'Tampax Towers' or the 'Toilet Rolls'.

In the Malay quarter are found buildings reminiscent of Dutch occupation. The houses in this area are simple and symmetrical with modest decorations. Malay immigrants built houses with locally available materials and the proximity of the houses to each other created a rich communal life.

A typical architectural style in Cape Town is known as Cape Dutch. Plan forms are rectangular and often symmetrical, usually with an H, T, U or E layout. One striking feature is the gables, which are very practical and designed to protect the house from fires and Southeaster storms. The single room (*dak kamer*) at top of the two-storied houses has a window in the gable from which the Dutch could see the ships in Table Bay (de Bosdari 1964: 24).

The architectural style changed with the passing of the Cape to British hands. The Cape Dutch houses were converted to suit the Georgian style, as the Britons replaced them with an architecture that fitted their own way of life.

British rule

The British view of Cape Town was that of an uncivilized, outlying district. But conversely Cape Town was also seen as a utopia, as the place of domination by the British Empire and the spreading of civilized culture. This last view is very clearly shown in a painting by the British artist James Ford in 1899, entitled 'Holiday in Cape Town in the Twentieth Century'. The painting shows Cape Town in a mixture of fact and fiction. There is an eclectic collection of buildings that actually existed in Cape Town, like the St Georges' Anglican Cathedral, buildings from abroad like the Eiffel Tower, and buildings that come directly from the imagination of the painter based on earlier architectural styles. The Cape Dutch style is totally absent in the painting and only two buildings refer to Dutch colonial rule. A rainbow that symbolizes the hope of a promising future occupies a very dominant place. The utopian view and the absence of Dutch elements can be explained by the time the painting was made: just before the Boer War in 1899, a time of optimistic British expansion.

The enormous British expansion coupled with their civic pride led to the erection of many monuments, memorials and portraits in bronze and stone expressing the domination of their culture. The city has many statues of British victors and notables, such as the former Governor of the Cape, Sir George Grey. There are two larger-than-life statues of Cecil John Rhodes, one of the greatest victors of the city. Other statues depict such leaders as Jan van Riebeeck, Louis Botha and Jan Smuts. The last two were Boer generals who were defeated but later became prime minister of the Union of South Africa. The statue of Botha is positioned in front of the Parliament Buildings.

To maximize their impact, statues were erected at important sites. Most of them were placed on an axis, which is an expression of the principle of order out of chaos. Compared to metropolises in Europe, America and the Far East, South African cities have enough room over for public statues in the city centres. There are several reasons for the comparatively low number of statues, namely the lack of state funding for such projects, the restricted budgets of museums and the relatively short history of most South African cities. Fifteen statues are situated in the Companies Garden that can be divided in two areas: the Deville Wood Memorial Garden and the Old Company's Garden. The Deville Wood Memorial, located in the former area, refers to the battle of Deville Wood in France in 1916. It is a symbol for the union of Afrikaner and English South Africans who fought together in France, and this is expressed in two figures who clasp hands across the back of a horse. The bronze sculpture was executed in 1925 as a replica of the central section of the memorial in France, erected near the site of the battle of Deville Wood. Another statue in this part of the garden reminds us of the men who fell in the First World War. These statues are state statues or following Crump and Van Niekerk (1988: 14): 'these war memorials are heraldic glorifications of a horror now imbued with hope, triumph and victory'.

In the Old Company's Garden it is the colonial period which is represented. Here we find statues of Queen Victoria, Jan Smuts (Figure 3), Governor Sir George Grey, and Cecil John Rhodes. The last was a politician, a businessman and an imperialist with large influence in South Africa. He died in 1902, dishonoured after an abortive attempt to gain control over the Transvaal Republic. Rhodes was devoured by an immense ambition and dreamt of an Anglo-Africa from Cairo to Cape Town. His statue in the Old Company's Garden raises his hand over Table Bay, symbolizing his ambition, something of which most Capetonians are aware. The inscription on this statue reads: 'Cecil John Rhodes 1853–1902 – your hinterland is there'.

The other statue of Rhodes is located on the Upper Campus of the University of Cape Town. This statue is related to the national level as well as to the local institutional level, namely the University of Cape Town. One and a half times life-size, the bronze statue of Rhodes, cast in 1934, faces north and has Table Mountain as its background. Cecil John Rhodes is sitting on a wall with his chin in the palm of his right hand and his right elbow on his thigh. His ambition is expressed in the inscription of the statue: 'I dream a dream – by rock and heath and pine – of empire to the northward – ay one land – from the lions head to the line'. Until Zimbabwe became independent, a replica of Rhodes' statue stood in Salisbury (Harare). Partially succeeding in the realization of his dream, Rhodes conquered Zimbabwe and Zambia. Both the statue in Cape Town and the statue in Salisbury faced northwards, emphasizing his dream. Because of the existence of a replica the statue gains more power.

The statue of Rhodes in Cape Town stands on an axis, namely the one that runs from below the University campus to the top of Devil's Peak (Phillips 1993). Originally, the statue did not stand on this line. In 1962, it was put there, as it had to give way to the construction of an underground at its original location. Now Rhodes also stands near the University, which emphasizes his importance for this place. He established the College, and the University was built on part of his estate.

Before its removal in 1962 there was a fence around the statue, but now the enclosure has been taken away and the statue has become easily accessible to the public. Students who pass it every day on their way to the campus still seem to notice it. They know who Rhodes was and refer to the Rhodes memorial, which is nearby. Some of them manage to mention the other statue of Rhodes in the city.

Figure 3 Statue of Jan Smuts (Photograph: Margot te Velde, 2004).

The establishment of the University was in line with Rhodes' ambition for a union of Africa. He thought students could be educated in that spirit. So, the university was linked to unionization and that concept is still valuable, albeit in a different sense, not as an ambition to create a union of Africa from Cairo to the Cape, but in the creation of a 'rainbow nation', a united South Africa in which all kinds of people live together.

Over the years the statue has been used by several groups. Students once painted it pink, expressing resentment. In 1996, a clothing shop advertised its jeans by putting them on the statue, with a board saying: 'Anyone can look cool in jeans'.

Apartheid

The years of apartheid have led to the creation of several important symbolic places and activities in Cape Town. The most evident are District Six, Robben Island, the Grand Parade and Coon Carnival.

The period of apartheid has left a visible scar on Cape Town. In the heart of the city lies a desolate wasteland, a symbol of the struggle and past oppressions. This used to be District Six: a racially mixed, highly overcrowded area, relatively poor, but with a very special, traditional character of its own. To its Malay inhabitants, District Six was known as Kanala Dorp. This came from the Malay word meaning to do a favour, literally 'if you please'. This was because a large number of the older houses were put up by groups of friends working for each other. This was an indication of the spirit of tolerance and community District Six came to symbolize.

The District was located in the close proximity to the city centre on the lower slopes of Devil's Peak, with views up the mountain and down to the harbour. The area consisted of small rectangular blocks, with the streets being the main place of work, of buying and selling, of meeting each other and of cultural ceremonies, such as the annual Coon Festival. The multicoloured houses were versions of the flat-roofed Cape Dutch houses, with British-influenced details such as the multi-panelled doors, recessed sash windows with larger panes and sliding upper halves (Fransen 1996: 19). 'The District represented non-racial, cosmopolitan living, everything the apartheid regime feared – and it was also a prime piece of land worth stealing,' one of the ex-residents said.

By the beginning of the twentieth century, the history of removals and segregation had begun. The first to be resettled were the blacks, forcibly removed from the District in 1901 after an outbreak of bubonic plague.

The exact population of District Six is unknown. In 1964 an estimated 61,000 coloured and Malay inhabitants, 800 whites and 600 Indians lived here. Two years later the major portion of District Six was proclaimed an area for white occupation and ownership in terms of the Group Areas Act of 1966. This proclamation was the cause of much repression. It was reported that residents in the area staged silent protests by walking through the streets holding notices saying 'I am from District Six', and by hanging black flags from their windows. But, do what they would, they would gradually and systematically be removed over a period of ten years. By 1982, all the black inhabitants had been forced to move to the then barren Cape Flats, on the other side of Table Mountain. Bulldozers flattened their houses. The city and the mountain were taken away from them. In an attempt to erase District Six from the map of Cape Town, the apartheid state tried to redesign the space of District Six by renaming it Zonnebloem.

Today, this part of town remains largely undeveloped. It is an area of wasteland with a few mosques and churches, the only surviving buildings. One of the churches, the former Methodist Chapel on Buitenkant Street, is now the District Six Museum. In its mission statement the District Six Museum project says: 'The area is still a powerful reminder of things that happened and must never be allowed to happen again. The District Six

Museum ensures this memory will not diminish.'

This Buitenkant Street church is a symbol of resistance and a sign of hope. The inscription on the plaque on the front of the church building reads: 'All who pass by, Remember with shame the many thousands of people who lived for generations in District Six and other parts of this city, and were forced to leave their homes because of the colour of their skins. Father, forgive us …'. Former residents writing their names where their houses once stood have filled in the museum's floor map. The museum also has a collection of District Six's original street signs. The contractor, who demolished the houses, kept the street signs and donated them to the museum in 1994, at the beginning of the new South Africa.

The land in District Six has never been fully redeveloped, as if the ground itself has been cursed. Given its symbolic status as well as the sorrow and anger of the victims of the forced removals, it is unavoidable that the redevelopment of the area is highly controversial (Le Grange 1996: 13). An ex-resident said: 'The land belongs to the people and only the people have to decide what to do with it. They cannot rob us again of our rights and our land.' The Restitution of Land Rights Act of 1994 and the Constitution form the legal framework for the resolution of land claims against the state. The Land Claims Court is responsible for deciding the validity of the claims. This is a very complicated and difficult process. There are many issues to be dealt with, which makes this process a long-term one. The location of former District Six is very central in Cape Town, which means that the market value of the land is very high. There are also multiple overlapping claims in respect of individual properties involving original owners, long-term tenants and even sub-tenants. Compensation for the emotional and economic trauma of the forced removals is another issue that needs to be addressed. People still feel very strongly about District Six. District Six can certainly not be redeveloped to what it formerly was, and its future remains uncertain.

Since it began to be bulldozed to the ground, District Six has been a favourite topic of local history. Many Capetonians knew the story of this part of town quite well. Upon being asked 'Can you tell me anything about District Six?' they described the history, the community, the annual Coon Carnival and the music. 'Lots of famous Coloured people came from District Six,' one person said. Life in District Six has been romanticized in musicals, poetry, paintings and so forth. Many ex-residents have written autobiographies about the time they lived in District Six.[3]

---

[3] See for example Rive (1986) and Forbes (1992).

D'Arcy has made a painting called 'The Devil's Revenge' (1988), which is a depiction of a District Six version of the legend of Van Hunks and the Devil, mentioned above. In 1996 at an exhibition in the South African National Gallery, the painter added the following text: 'Deep in the heart of incomparable Cape Town lies a large tract of fallow land, an urban scar that defiles the landscape like a pustule sore on a queen's forehead. It's a cursed piece of earth, soaked in tears, yet hankered after like a Jewish Jerusalem, a Muslim Mecca.'

The painting tells how the Devil, after having lost the bet to Van Hunks, swore revenge, promising to inflict misery on the area below the mountain now known as District Six. For six generations the Devil plotted his evil, until, eventually, in the seventh, he manipulated the mind of the members of Parliament to pass the Group Areas Act in 1966. The painting shows the Devil on top of Table Mountain smiling sardonically at the drama of destruction below. A central image in the painting is a white masked driver, scooping residents – offspring of Van Hunks – to the Cape Flats. Another naked person is shown, back turned on his future, and with an ambivalent gesture of the hand – half a plea to stop, half a wave goodbye to the past – stripped bare of dignity (Maurice 1995: 15).

Besides District Six, Robben Island has assumed an important place in the symbolic landscape of Cape Town. This island lies in Table Bay about ten kilometres across from the Mouille Point lighthouse. Over the last 350 years this island has been used to isolate people from society. Among the prisoners and exiles have been Khoikhoi, holy Muslim leaders, sultans, black chiefs who resisted British colonial rule, lepers, lunatics, spiritualists and the leaders of the struggle against apartheid.

Internationally, this island is best known for the imprisonment of Nelson Mandela. At this level, Robben Island symbolizes the ability of South Africa to triumph over adversity and to achieve a peaceful and stable democratic society. This was emphasized during the first ever visit of a US President, Clinton, to South Africa at the beginning of 1998. He visited the prison cells on Robben Island together with the returning Mandela. In the media this was portrayed as a sign that Mandela and South Africa had won.

Nationally, the island has been seen as a place of banishment where the unwanted of society have been put away. The years in which the island was a maximum security political prison (1959–1991) have reinforced the symbolic value of the place for many South Africans as a site of repression and resistance.

Locally, Robben Island has a mystique borrowed from its isolation and distance from Cape Town. For many Robben Island prisoners, both political and criminal, the mainland and their motherland, was District Six.

At a ceremony in the Cape Town's Town Hall on 27 November 1997, Nelson Mandela was honoured with 'Freedom of the City of Cape Town', the highest award the City may give to an individual. The City Manager ended her citation to the award as follows: 'His long walk to freedom ended here in our city. The people of Cape Town recognize his struggle for non-racialism and democracy, his strength of spirit, his courage to endure and determination to achieve a dream. As a symbol of fortitude and hope, he stands as an inspiration to us all' (President 1997: 1).

Ahmed Kathrada, a former political prisoner and chairperson of the Future of Robben Island Committee, stated on Heritage Day, 1977: 'While we will not forget the brutality of apartheid we will not want Robben Island to be a monument to our hardship and suffering. We would want it to be a monument reflecting the triumph of the human spirit against the forces of evil; a triumph of freedom and human dignity over repression and humiliation; a triumph of wisdom and largeness of spirit against small minds and pettiness; a triumph of courage and determination over human frailty and weakness; a triumph of non-racialism over bigotry and intolerance; a triumph of the new South Africa over the old' (Department of Arts and Culture 14-8-1997).

Nowadays, Robben Island has evolved into a transition phase, because of international and commercial interests. New symbolism has developed in the form of putting a stone on a pile reminiscent to the stone that Nelson Mandela laid down at Robben Island on his first return after freedom. Other ex-prisoners have followed him and the mound has become an icon of freedom, while the act of adding a stone has become a symbol of hope (ter Berg 2004: note 102).

The apartheid years are best visualized in the urban arena in terms of spatial segregation. As seen above in District Six, a total separation policy was practised by forced removals, resulting in the formation of segregated townships. This was a very drastic form of segregation. A less clear form of separation is the use of public space in the city based on race, without the official imprint of segregation. This is still present in the Cape Town of today. Nowadays, however, this segregation that was once based primarily on race is also related to economic access.

There are certain places in the central district of Cape Town that are predominantly used by low-income blacks and coloureds, such as the Company's Garden, and the footpaths of Adderley Street. The main street of Cape Town is the location of street hawkers, people living in the Cape Flats. Another place in Cape Town which is mostly used by low-income blacks and coloured is the Grand Parade. In colonial days this square was used for public executions and called the 'Exercitie Plijn' and 'Wapen Plijn'. It is located in front of the Castle of Good Hope and surrounded by buildings such as the Town Hall and the Cape Town station. This railway complex is divided into two areas in accordance with the ideology of apartheid architecture.

Today in Cape Town, the informal sector represents a substantial proportion of the local economy. On Wednesday and Saturday the entire square is transformed into a marketplace, with clothing and food placed on portable trestle tables and plastic sheets. The goods are sold by and to people from the Cape Flats. Historically, this square has been used in the struggle for justice and freedom. Gatherings and celebrations in this public square in Cape Town showed the first black presences in the city. Protest meetings – stopped by armed police and troops – against the forced removals of the blacks from District Six were held in this square at the beginning of the twentieth century. There was a mass rally on the Grand Parade at the release of Nelson Mandela from prison in 1990. Nelson Mandela delivered a historic address from the Town Hall. It was also on this site that Mandela gave a speech to a large crowd the day he was elected State President on 9 May 1994.

The Coon Carnival, as mentioned above, is also important for the town's symbolism. Its origins lie years before the institution of apartheid. It was originally a celebration of the New Year among the Cape's slave community, incorporating the whole of the inner city. After 1834 and the abolition of slavery, it assumed more the guise of a freedom march. Several blacks participated, but the ethnic character of the carnival was coloured (Jeppie and Soudien 1990: 80). The early troupes were often called the *Kaapse Klopse* or Cape Clubs.

The performers were inspired by American minstrels visiting Cape Town in the late nineteenth century and copied their looks: faces covered with black make-up and eyes encircled in white like racoons (the North American animal). They dressed in striped blazers and white flannel trousers and Panama hats. The music performed was a blend of Dutch and Eastern music. The Cape Malay fisherman and the Dutch sailors exchanged songs in the Bay, which led to a unique combined musical culture. The songs are called *ghommaliedjies*, derived from the word *ghomma*, the Cape's ethnic bongo drum imported from the East. It is used in combination with guitars and tambourines.

Until 1968 the carnival parade route traced the boundaries of the areas where the Coons lived (*Cape Times* 27-1-1998). District Six was the heart of the Coon Carnival. Major routes such as Hanover Street and Constitution Street in District Six were part of the route of the annual carnival marches, until the forced removals of 1966 and the Gatherings Act of 1973. This law restricted the route of the Coons. They were forced to parade off the Cape Flats. The carnival has survived, despite the apartheid regulations. Many coloured political activists condemned the Coons of the apartheid years for playing up the racist stereotypes of stupid, singing slaves. But those involved insisted it was a celebration of freedom, not oppression, and that their disguise made fun of the racist stereotype rather than reinforcing it (*Cape Argus* 3-1-1997).

Today the annual Coon Carnival is a formal competition between rival troupes gathered in stadiums. It is no longer held in the streets of the city where it really belongs. The *ghommaliedjies* are performed for tourists and Capetonians on Cape Town's popular beaches during the summer season. This historical celebration of inner-city history and culture has been commercialized and modernized, but it remains a public reminder of a past, a celebration of community, survival and liberation.

Transition and modernization

The state has no coherent national policy to commemorate anti-apartheid heroes by means of statues and the like. The Truth and Reconciliation Commission, which advises the government on reparation policy, has already indicated that monuments should form part of thinking about reparation.

Outside the East London Town Hall in September 1997, a larger-than-life brass statue of 'hero of the nation' Steven Biko was the first statue of an anti-apartheid leader to be put up in South Africa, after having been commissioned by the Biko family and a newspaper editor and funded by international film stars. The black mayor said at the unveiling: 'The statue will be a constant reminder to us that there was tyranny. And there were people who were good enough to pay sacrifices, to the extent of being incarcerated and laying down their lives.' However, the successors to the Black Consciousness Movement, Azapo, expressed serious reservations about the fact that the commission chose a white sculptor. They would have preferred a black sculptor for 'empowerment'. Members of the crowd during the unveiling were interviewed and none of those believed the statue was a good likeness of the 31-year-old activist, who died in 1977.

In October 1997, a number of statues in Pretoria and Johannesburg were draped in black cloths as part of the stand of the magazine *Tribute* against public artworks of the old South Africa that meant nothing to the vast majority of people. 'Monuments that have stood for ages, erected to men who represent all we have struggled to change, are a little darkened this morning. They have been transformed into billboards of visual justice,' a spokesperson of the magazine said (*Cape Times* 27-10-1997).

At present, in Cape Town there are no statues commemorating the black or coloured in the country. Nor are there statues made by blacks or coloureds, and the City Council does not specifically look for black or coloured artists for possible new statues. Their historical culture and individual leaders, such as Nelson Mandela, have not yet been recognized or commemorated by the present government in this form. The city landscape is still dominated by images of white men. There are no statues to writers, poets, artists or philosophers.

None of the main informants showed any real interest in symbolizing the post-apartheid years through statues. Comments such as 'we are too busy focusing on our daily work to think about things like that' and 'we should rather spend money on more important issues as crime and improvement of slums' were frequently heard when asked if they thought there should be more statues to commemorate the anti-apartheid leaders.

The City Council does have plans to erect a World Peace Monument similar to the one in New Zealand on council-owned land in the city, preferably in a park, in the near future. It has recently put an application forward to all Cape Town residents for suggestions for a site for construction or to hear any objections to the application. The monument will be very large, about eight metres high on an eight-metre base. It will be the first of its kind in Africa and is to be built by the highest monk in Burma, named Sayadaw, who is 90 years of age. He has built 770 monuments throughout Burma and twenty-three World Peace Monuments internationally, on all continents except Africa.

There were plans in the City Council for a statue of a hand of Mandela breaking through prison bars, but nothing came of this proposal, mostly because of a lack of funding. This abortive project is just one of many examples. The City Council currently has plans for a memorial park in District Six, in which a part of the Berlin Wall may be placed.

New sculptures will be a result of competitions, without special themes,[4] and organized by the Arts and Culture Committee with commitment by the City Council. Location ideas for new sculptures are: at termination of axes; in pedestrian areas; at lost or forgotten historical places; at places where they add to the experience of the specific place; and at new development areas

---

[4] Although one city planner talked about a possible proposal for the theme 'transformation'.

(Inner-city Planning Unit, City Council).

One sculpture, much talked about, that has been located temporarily in the Waterfront area was a thirty-metre-high statue of pop star Michael Jackson. The immense fibreglass statue was flown in before the concert given during the singer's visit to South Africa in November 1997, and transported to Johannesburg after the performance. It was Michael Jackson's first ever tour in South Africa after the apartheid years. Many black Capetonians see him as a hero and his visit as a symbol of world acceptance of South Africa.

The most recently erected sculpture in the city is in the form of an ice cream, some two metres high. This sculpture, made by Cevin Young from the Fine Arts Department of the University of Cape Town, is placed on the Waterfront, in front of an American franchise restaurant 'Planet Hollywood', in an area of new development. None of the fifteen main informants knew where the sculpture was located after being shown a picture of it. They had never seen it before. This sculpture is a symbol of globalization and modernization located in the most modern part of the city.

Traditional dancing, poetry, storytelling, painting and performing arts have always been part of South African life. They form a rich cultural heritage ignored during the apartheid years. This situation is changing.

Cape Town is now graced by several murals. There are many, smaller murals in Cape Town, made by individuals or groups. Most are painted on fences surrounding school grounds and slum areas (in the Cape Flats paintings around squatter camps are especially numerous). They are often paintings created by children and represent happy, colourful images, for example of boats, of the sun, a rainbow, Table Mountain, or of children playing. The rainbow nation is symbolized in many paintings. Children's paintings around school grounds form the fence of a peaceful and innocent area, safeguarding the children from the outside world. The murals symbolize hope for the future, innocence and peace.

A striking and very colourful mural is displayed on a building facing Stal Plijn outside Parliament in the centre of the city. It measures seventeen by twelve metres, which makes it visible even from the top of Signal Hill. The idea of this mural was proposed by Adli Jacobs and Abubakr Shabudien and approved by the National Constitutional Assembly. This 'official' mural is a depiction of the first national democratic election in 1994 (Figure 4). It was displayed during a ceremony when the final constitution was adopted in 1996, addressed by President Mandela.

The purpose of this mural is to monumentalize the celebration and accord proper recognition to the political and historical ritual of the 1994 elections.

The depiction is a representation of the transition in South Africa. It is a way to show the struggle for freedom and democracy as well as the building of the rainbow nation.

The location of the mural is very strategic, in the middle of the city, facing Signal Hill, the statue of General Louis Botha and the Parliament buildings. The Parliament buildings – an imposing example of Victorian and Edwardian architecture – were once a symbol of prejudice and intolerance, but have now become a symbol of hope for the future of the country.

The mural is made of four giant separate canvases, each with its own theme. They can be read separately or together, like a picture book. If we look at the mural the four sub-themes are from lower left, lower right, top left to top right, we see:

The Struggle: On the canvas we see fire, darkness, guns, protesting people, a bare landscape, an empty school and dark houses separated by a large fence.

Negotiations and Interim Constitution: Symbolized by a ballot box dated 27 April 1994; a negotiating table with people sitting around it, people carrying signs as 'your vote is secret' and 'we want an accountable government', a mother carrying her child, and a rising sun.

Democratic elections: In this part of the mural there is a government house, a black leader, the first glimmerings of a rainbow, black and white people talking to each other and a green tree, which is a symbol for peace and of confidence in the future and of life and hope according to F.W. de Klerk (1994).

The New Constitution and the Future: In the most colourful canvas we find symbols for peace, hope, unification and prosperity such as the white dove and the tree in bloom, children of all races walking to school together, a flourishing industry, people holding hands, houses without fences, water, a fertile land, the new South African flag, a shining sun and, of course, the rainbow.

The symbolism used in the painting is simple and the colours are those of the rainbow. The bottom half is darker than the top half.

This mural is one of few examples of recent monumentalization in Cape Town of the country's apartheid years and its struggle and transition. Some persons had not noticed it, in spite of its size. For them it had been swallowed up in the urban background. Others could tell precisely where to find it. Being asked about its meaning they gave such answers as 'it is about who to vote for in the rainbow nation' and 'if you vote it will be a happy nation'.

Figure 4 Mural (Photograph: Margot te Velde, 2004).

In the past ten years, Cape Town's Foreshore has been turned into a site for commercial development. Here one finds modern shopping malls with up-market international stores. The seven-year-old Victoria and Alfred Waterfront Harbour complex is believed to be Africa's largest single entertainment attraction. More than twenty million people visited the highly successful V&A Waterfront complex in 1997, a 12 per cent increase on 1996 (*Cape Times* 12-1-1998).

In 1994 in this area a series of thirty-two historical panels was installed by the Waterfront Company that is in charge of the development. The panels were produced by historians at the University of Cape Town and show the port's history. The text and visual images trace the history of the racially diverse working population, including slave labour, in the shipping and construction industries of the port. One of the panels focuses on nearby Robben Island with the following text: 'One of the world's most notorious political prisons where Nelson Mandela spent years for resisting apartheid'.

The panels reflect the fact that the 'new history' has commercial appeal. Frankly acknowledging the troubled past makes the area seem up-to-date, in tune with the spirit of post-apartheid South Africa (Foner 1995: 164).

The Foreshore and the city centre are now connected by two freeways. In the 1980s, city planners and engineers had designed a more extended highway system with several freeways and intersections. A hitch arose during the construction of the connection between the Foreshore and the City Bowl, when state money run out. The result is a very interesting site, namely an intersection and two open-ended freeways on each side of the intersection that were supposed to connect above it (Figure 5). In the next few years the ends may be connected. This failure enjoys plenty of local notoriety; every informant knew where this uncompleted freeway was located, after being shown a picture of it. It has also been the location of several television commercials.

It is symbolic of the past (local) government and its failures, but also of the (slow) transition. During the Cape Town Olympic Bid in 1997, someone painted a text on the lower part of the freeway saying: 'If we win the Olympic bid, we can finish the bridge and finish unfinished things.' Cape Town did not win the 2004 Olympic bid, and the future of the intersection remains uncertain.

Ten years of democracy

The year 2004 marked the tenth year of democracy in South Africa and the ANC has won the Western Cape Province election for the first time in history.

In February 2002, the city of Pietersburg was changed to Polokwane and Potgietersrus to Mokopane in the Northern Province, now Limpopo. A political row occurred in Cape Town in 2001 when the then mayor announced his plans to rename Adderley and Wale Streets. The initiative ended in public hearings and allegations of vote rigging which caused a shift in political power in the province (*Cape Times* 3-4-2002). In 2003 the Cape Town City Council drew up a new policy for renaming streets, but the

Figure 5 Open-ended freeway (Photograph: Margot te Velde, 2004).

general idea is that streets should only be renamed in exceptional circumstances, in order not to cause racial divisions or sectoral bitterness.Any person, community or organization within the boundaries of the city of Cape Town is entitled to propose the renaming of a street, after which a lengthy process of reviewing the application will begin. Among the criteria in the report for new name selection are (1) to honour or commemorate noteworthy persons associated with the city and who have contributed significantly to its welfare and that of its citizens and (2) to recognize native wildlife, flora, fauna or natural features related to the community and city. To date, no street names have been changed.

The city drew up an action plan for the conservation and management of the Company's Garden in 2002, after a public study of the symbolic, natural and cultural values of the landscape. This was done after public pressure, primarily due to the deterioration of the Garden. Until today there have not been any significant improvements.

A statue of Nelson Mandela was erected in Johannesburg and a statue of Freedom (a 110 metre stature of Nelson Mandela) has been planned for the harbour mouth of Port Elizabeth. In Cape Town, however, no new statues have been erected and there are no firm plans for the near future.

In February 2004, a 'Homecoming of the Elders' ceremony was held for the elderly former residents of District Six. The international media covered the ceremony, and Nelson Mandela was present. Nine semi-detached houses in Chapel Street in District Six were rebuilt. However, the 'homecoming' event was merely symbolic, as the houses were only nearing completion because the trust that represents the returning residents is strapped for cash. The event was organized by the trust to coincide with the anniversary of the District Six being declared a white group area.[5]

Robben Island has been connected to the shores with the opening of the Robben Island Museum in the Robben Island Gateway at the Waterfront. Chairman of the Robben Island Museum, Ahmed Kathrada, stated that the Robben Island Gateway building would provide Cape Town with a tangible symbol of the importance of the island for South Africa's new democracy.

The Waterfront area has been strongly developed into up-market, residential and shopping areas, including the Clock Tower section. The original Clock Tower itself is a national monument dating back to 1882 when it was used as the Port Captain's headquarters. Alongside the restored Clock Tower are the ruins of the Chavonnes Battery, a military installation from the Dutch East India Company era.[6]

Canals have been re-dug from the Table Bay to connect the newly opened International Convention Centre on the Foreshore. This ultra-modern piece of architecture is both liked or strongly disliked by Capetonians, based on the numerous sent in letters to the newspapers.

Next to the Convention Centre, we still find the open-ended freeways on each side of the intersection, but possibly not for long, as South Africa – including Cape Town – has finally won the bid for the 2010 Soccer World Cup. Perhaps this event will bring the reconciliation needed to fill the symbolic void in Cape Town; a unity which is already found in the rest of South Africa.

Conclusion

Cape Town's symbolism is very diverse, involving urban layout, specific places, statues, murals, architecture, urban myths and festivities. Predominant, however, is the natural environment, consisting of Table Mountain, Devil's Peak, Lion's Head and Table Bay. The superlative, ascendant natural beauty dominates the city's cultural aspect, enforcing a

[5] See Gill Moodie, *Sunday Times* 16 May 2004.
[6] Source: www.waterfront.co.za/clocktower/history.php

strong continuity throughout history. This is a very unique trait of Cape Town's symbolism that is generally not found in other cities.

The natural beauty has meaning as a spontaneous symbol in contrast to the many created symbols present in the city. Because of the strong dominance of nature as a spontaneous symbol accepted by all groups in urban society, in phase of transition this natural monument may substitute symbols that otherwise would be necessary to create a feeling of togetherness. The city's urban symbolism also reflects social change in its layering according to the different periods related to the various population groups in power. The symbolism of these different periods is very clear and crystallized. This, however, does not hold for the post-apartheid period. Some officials say that symbolism is less important nowadays as other more pressing issues such as crime and housing have to be addressed.

These problems need not be brushed off by the observation that urban symbolism in the post-apartheid era has raised its own questions to be addressed. Given the many abortive plans and proposals for new symbolism, it is obvious that the government and population are not ready for it. In our opinion an important factor could be that President Mandela did not have a major inclination for furtherance of urban symbolism. He – in contrast to Sukarno as the main director of urban symbolic change in Jakarta – did not aspire to change the symbolic structure of the town in the framework of nation building, although it has to be pointed out that the concept and practice of the National Truth Commission has got strong international symbolic power. At the present moment no symbols, except perhaps Nelson Mandela himself, are acceptable to all the population groups in Cape Town. Urban community spirit in the post-apartheid period has not evolved far enough to settle the disputes on symbolism by a consensus about specific symbols. Urban symbolism can be used as a strategy for the advancement of urban communal spirit, but to reach that objective a minimum of collective support is needed. This requirement has not yet been met, and the filling of Cape Town's post-apartheid 'symbolic void' will demand aeons of time and plenty of intensive debate, although it is still supported by a strong layer of natural beauty.

References

Berg, R. ter (2004) *The City as Text: Images, Space and Symbols in Cape Town, South Africa*. Leiden. [MA thesis, Leiden University.]

Bosdari, C. de (1964) *Dutch Houses and Farms: Their Architecture and History Together with a Note of the Role of Cecil John Rhodes in their Preservation*. Cape Town: Balkema.

Cape Argus.

Cape Times.

Crump, A. and R. van Niekerk (1988) *Public Statues and Reliefs of Central Cape Town.* Cape Town: SA Gallery Press.

Department of Arts and Culture (14-8-1997) *Statement by the Department of Arts and Culture, Science and Technology.* [www.gcis.gov.za/cgi-bin/vdkw_cgi/xb536916e-24/search/4892588/30]

Fagan, G. (1985) *The Company's Garden: Report for the Cape Town City Council.* Cape Town: City Council.

Foner, E. (1995) 'We must forget the past: History in the new South Africa.' *South African Historical Journal*, vol. 32, pp. 163–76.

Forbes, A. (1992) *Birds on a Ledge.* Cape Town: Buchu Books.

Fransen, H. (1996) 'The architecture that Cape Town lost.' In: *The last days of District Six*, pp. 19–23. Cape Town: The District Six Museum Foundation.

Grange, L. Le (1996) 'The urbanism of District Six.' In: *The Last Days of District Six*, pp. 7–17. Cape Town: The District Six Museum Foundation.

Jeppie, S. and C. Soudien (eds) (1990) *The Struggle for District Six: Past and Present.* Cape Town: Buchu Books.

Klerk, F.W. de (2-9-1994) Speech of the Deputy President during the planting of a peace tree on the Grand Parade as part of the arbour day celebrations. [www.gcis.gov.za/cgi-bin/vdkw_cgi/sb609a763-10/Search/5033521/6]

Kostof, S. (1991) *The City Shaped: Urban Patterns and Meaning through History.* London: Thames and Hudson.

Maurice, E. (1995) 'The sore on the queen's forehead.' In: *District Six: Image and Representation.* Cape Town: South African National Gallery.

Miller, P. (1979) *Myths and Legends of Southern Africa.* Cape Town: T.V. Bulpin Publications.

Philips, H. (1993) *The University of Cape Town 1918-1948: Formative Years.* Cape Town: UCT Press.

President (1997) President Nelson Mandela receives the city's highest honour. In: *City of Cape Town Bulletin*, vol. 19, no. 12.

Rive, R. (1986) *Buckingham Palace: District Six.* Cape Town: David Philip.

Southall, A. (1993) 'The circle and the square: Symbolic form and process in the city.' In: P.J.M. Nas (ed.), *Urban Symbolism*, pp. 378–92. Leiden: E.J. Brill.

Sunday Times.

Velde, M.N. te (1998) *Between Devil's Peak and Lion's Head: Urban Symbolism in Cape Town, South Africa.* Leiden. [MA thesis, Leiden University.]

Chapter 8

THE CONTRA-DANCE
Rapping and the art of symbolic resistance
in the Parisian *banlieue*

Leeke Reinders

Introduction

The study of urban symbols points us to the ways people comprehend, bring order to and make sense of their existence. Urban life, having a tendency to be a fragmented and chaotic experience, is thus rendered clear. Symbols then relate to a sense of order in fragmentation. However, the symbolic urban arena is not a fixed but a contested order. 'The symbolic order,' as Peter Nas proclaimed in his introductory work, 'often proves to be ambiguous, amorphous, fragmented and incoherent, a symbolic configuration of hybrid and shifting deconstructed images arisen from tensions, conflicts and social changes' (Nas 1993: 5–6). The study of urban symbolism has concerned itself with a whole terrain of symbolic forms, but until now has largely focused on the way symbols materialize into the physical environment, as in statues and monuments. Below, attention is shifted from visual and material symbolizations to a less tangible but nonetheless powerful form of urban symbolism. This chapter deals with the social and textual practice of rapping in Sarcelles, a degraded suburb some fifteen kilometres northeast of Paris. Sarcelles is located in the *banlieue*, an area caught in the popular imagination as an area which is a concentration of social problems, varying from unemployment to drug dealing and petty crime to prostitution and xenophobia. Youths form a dominant grouping in the city, not only as the category – up to twenty-years-old – forms the biggest age group, but also in the emphatic manner they make their presence felt in the city. *Les beurs*, as these migrant kids are also known, use obscene words and gestures, go around dressed in trainers and tracksuits, cluster together at suspicious places, write on walls, and make offensive rap noises. In contrast with what is often assumed, these forms of expression are no empty poses or sloganeering. Below, making use of the notion of the 'hidden transcript', we see how young men create a contra-symbolic discourse, in which they claim their dominance. In the first section, some guiding ideas on the social text of rap are formulated, with reference to notions of exclusion and marginality, which dominate the public debate on young people in the *banlieue*. In the sections that follow several principles of the symbolic language of rap are sketched out. Here, we see how groups of young men use rap as social weaponry, in which they seek the dignity and reputation regular society denies them.[1]

Symbols, text and the hidden transcript of rap

A warm wind wafts across the room. Azzdine sits in the back, his arms folded and legs casually stretched out on the floor. He pricks up his ear and nods approvingly to a monotonous bass sound, coming from the home stereo system. The walls behind him are covered in posters of American pop and movie stars. The floor is filled with bottles, ashtrays and other remnants of yesterday's party.

At the other end of the room Achmed leans back on a worn-out bench. For a long time he stays aloof, listening with half an ear to the stereo. After a while he straightens his back and takes out a creased paper from his back pocket. Holding it tightly he hesitatingly starts to read from the paper. In the lyric he describes an unfolding illegal drug deal. 'With all this money ahead you want to drop everything,' he reads. 'I settle by spending.' Achmed counts his blessings as the stereo continues. Halfway through the lyric Filipe joins in. He speaks from memory, his eyes fixed on the window. After the second refrain host Mario takes command and a triangle is formed, in which the three rappers each show off their vocal capacities. With stretched-out thumbs and little fingers they accentuate witty lines, puns and other critical passages in the text. Suddenly Achmed, Filipe and Mario pause. A friend just came in.

---

[1] This article is based on an ethnographic study on the relation between place and identity in the Parisian *banlieue* of Sarcelles, carried out in the summers of 1997 and 2001. An earlier version of this chapter was presented at the international workshop 'Urban Symbolism' in Leiden, June 16–18, 2004. The author wishes to thank the participants in the workshop for their constructive comments. A warm word of thanks goes out to Rivke Jaffe, who translated my Dutch transcriptions of the idiosyncratic language of French rappers into legible and 'street credible' English.

'How did it go,' Mario asks.

'They fucked me,' the boy answers. He had just finished his driving exam and failed. 'One wrong move. That did it.'

While everyone listens to the account of the exam, Azzdine leaves. Filipe, a twenty-one-year-old motor mechanic, recounts his experiences with driving exams and the conversation switches to automobile brands. The boys exchange ideas on the point of having exams, the SIMCA parked in front of the house, the price of second-hand cars and slowly the conversation dissolves into the rhythm of the programmed computer beat. With the rhythm of speaking and the tone of their voices, the conversation somehow seems to blend into the rap song that started ten minutes ago.

During the meeting a noteworthy thing happened. In spite of the catchy rhythm and the playful way they performed their raps, the boys kept almost motionless. They would shake their heads or twist their arms, but nobody tapped his feet. 'Rap is not for dancing,' another rapper, Thiburce, told me later on. Thiburce's features spoke volumes. Once in a while during our conversations he would act one of his lyrics. At a glance his mimicry would change. His eyes, before all friendly and approachable, spoke fire and his face showed a grimace. This seems to be pointless, but actually has significance. Rap is serious business. The practice of rapping originally refers to a form of street talk or everyday conversation carried on in face-to-face exchanges by which people could establish and maintain a public reputation (Abrahams 1974; Kochman 1972). As a way of public speaking, originating in Afro-American culture, rapping has many antecedents. In the Southern states forms of verbal competition were enacted 'by the dozens', in which group members exchanged insults. In the Caribbean community of Harlem, New York, reggae toasting came into existence, a musical form strongly influenced by the 'jive-talking' or quick talking of disc jockeys. 'Toasts' are narrative poems, in which men tell each other obscene and violent stories to get through times of forced boredom, as in prison, in the army or on the street. Similar verbal forms are found in the practice of 'scatting', in which performers use their voice as an instrument in games of question-and-answer with the audience. And 'soul rap' elaborated on spiritual sermons in Afro-American churches, in which the verbal reactions of the community played an important part. These ways of talking can be traced into musical forms such as disco, bebop, doo-wop, a cappella groups, street funk and stand-up comedy, and all point to a recurrent motive: the social importance of a glib tongue.

During the 1970s rap, alongside the practices of break-dancing and graffiti, became an integral part of hip-hop culture. In the New York Bronx young Latins, Afro-Americans and Afro-Caribbeans used rap as a substitute for the gang wars which terrorized the neighbourhood (Toop 1984). In the decades that followed, rap underwent severe changes. Originated as a competitive narrative form, embedded in everyday life and face-to-face contacts and interactions, rap became part of a popular culture. Rap was launched, for example, as a medium to advocate a Black Nationalist consciousness and to create solidarity beyond the boundaries of the local community (Mitchell 1998). In common with the development of other 'black' musical styles, such as jazz, soul, funk and gospel, rap also turned from a local cultural practice into a mass cultural form. During this process of commercialization rap often received a more marketable and less radical form of expression (Blair 1993). A sharp division between public and performer was created, replacing the spontaneity of live production and performance, characteristic of the original social practice of rapping. Rap, as it turned out, largely went out of the public domain of the street and into the private sphere of Walkmans and home stereos (Dimitriadis 1996). These processes of globalization and commodification show rap to be a highly flexible and adaptive social and musical genre, easily combined with different idioms and adopted in various settings.

Below, attention is focused on the way young people localize a global form of expression and appropriate rap as an alternative means of communication, whereby they wrest themselves free from a marginal existence in a peripheral place. The place of action is the new town of Sarcelles, located fifteen kilometres northeast of Paris. Sarcelles was built during the post-war years according to the principles of modernistic planning and architecture. The buildings that arose were made of modern materials such as steel and concrete, standardized production methods ensured rapid and cheap construction, and the rational design of the neighbourhoods was meant to mould the residents into a community. In the last three decades, however, Sarcelles has gained the reputation as an area with a concentration of social problems, ranging from petty crime and drug abuse to unemployment and xenophobia. Responsibility for the decline of the city is largely blamed on the second generation of youths who, in contrast with their parents, were born in the city and/or grew up there. 'Youths,' writes Wacquant, 'are widely singled out by older residents as the chief source of vandalism, delinquency, and insecurity, and they are publicly held up as responsible for the worsening condition and reputation of the degraded *banlieue*' (Wacquant 1993: 376). In this sense the *banlieue* differs from the case of the ghettoization of North American inner cities, since it is age, not ethnicity, which is considered to be the most important factor in the decline of the city.

Below, attention is focused on rap as a counter-discursive means of expression, in which young men in more or less subtle ways stylize and formulate their resistance against the established order. Rapping is seen as a symbolic form of expression. As defined by Clifford Geertz, a symbol is a 'tangible formulation of notions, abstractions from experience fixed in perceptible forms, concrete embodiments of ideas, attitudes, judgements, longings or beliefs' (Geertz 1973: 91). According to Geertz, symbols always relate to an 'intersubjective world of common understandings' (Geertz 1973: 92). The inherent discordant nature of symbols results from their multivocality. A symbol always represents many things at the same time. As Victor Turner writes: 'Its referents are not all of the same logical order but are drawn from many domains of social experience and ethical evaluation' (Turner 1982: 52). A study on urban symbolism thus gives insight into the working of power relations in society.

This paper starts with the notion of the 'hidden transcript', a concept derived from the work of James Scott (1985, 1990). According to Scott, social struggles not only have material effects and causes, but also centre on symbolism. 'The symbolic high ground between groups and classes will remain an integral part of any conflict over power' (Scott 1985: 235). In *Domination and the Arts of Resistance* Scott shows how power relations are caught up in an ambiguous discourse. There exists a discrepancy between what people say and do and what they think and mean. People translate these contradictions into different forms of social interaction, whereby social forms of subordination are publicly displayed. To analyze these subversive practices, Scott places two domains side by side. On the one hand, there exists the status quo of the 'public transcript'. By this he means the social rules, mainly enforced by those in authority, that make interactions between unequals run smoothly. However, things are not always what they seem to be. In the public arena, both dominant and subordinate groups make use of hidden signals. With the help of 'hidden transcripts' people disguise their intentions for the sake of the public order. 'The hidden transcript,' Scott writes, 'is the privileged site for non-hegemonic, contrapuntal, dissident, subversive discourse' (Scott 1990: 25). This resistance against the established order takes place in social sites. This can be a physical location, but also takes shape in linguistic codes, gestures and other forms of verbal and non-verbal behaviour. What is crucial here is that it concerns messages that both confirm and contradict the public transcript. The trick is to make these messages public in a veiled manner.

Hidden transcripts have their ideal breeding ground in sites which escape the eyes of the power holders. Off stage, taken out of a power-loaded situation, people turn to their congeners to express their views in relative freedom. Within the protected environment of a group, people seek revenge. Here, attempts to keep subordinates under one's thumb have a reciprocal resistance from below.

Although Scott writes about relatively serious forms of oppression and exploitation, such as slavery, racism and colonial rule, his ideas throw an enlightening view on the social practices of rapping in Sarcelles. The symbolic resistance of rappers rejects the categories the power holders attempt to impose on them, by confirming them in a twisted manner. The verbal play of rap, as a war of words, can be seen as a practice of what Scott refers to as 'everyday resistance'. In their rap lyrics we see different elements of the hidden transcript at work. Below, four principles of the hidden transcript of rapping are touched upon. Here, we see how by prescribing an ethos of solidarity, masking message and messenger, romanticizing their marginal position and seeking revenge, rappers construct a symbolic and textual universe, within which they form a united front against wider society. Rapping functions as social weaponry, provoking the publicly accepted and turning the established order upside down. Here young men find the self-confidence, dignity and reputation which regular society denies them. Within rap, a contest for symbolic authority can be discerned, which undercuts the moral and social authority of their enemies by allocating a resource over which they have some control: language. Rap, as shown in the above description of a rap session, offers young people an opportunity to exchange stories in a small circle of friends and thereby has a big potential. Although the practice of rap often involves accessories, such as records and musical equipment, in essence it has only one constituent: the human voice. In this sense rap is an easily accessible medium and the rapper an autodidact. After all, everyone with good working eardrums and vocal chords is able to rap. Moreover, the human voice is an adaptive medium and difficult to supervise. 'The human voice,' as Scott writes, 'is uncontrollable' (Scott 1990: 162).

Mythmaking and the ethos of solidarity

Rap lyrics are usually performed by several persons. Rapping is an interactive practice, in which group members by turns take passages of a text. Rappers often make use of the practice of *bouster*, whereby words are spoken by plural voices at regular intervals in the text. Figures 1a and 1b show two written lyrics from the notebooks of local rappers. The text by

Thiburce contains several notes written in the margin referring to extracts to be spoken by two members of his rap group. Filipe has encircled rhyming words, which are also to be spoken by several persons. This format of polyvocality is not only a musical instrumentation to accompany the rhythmic course of a rap song, but more importantly serves a social function. During a rap session young men exchange stories and thereby support each other in carrying out their messages. In this sense rapping is a specific form of sociability. 'Sociability,' to quote Ulf Hannerz, 'is a kind of interaction in which definitions and evaluations of self, others, and the external world are developed, maintained, and displayed with greater intensity than in other interaction.' 'Successful sociability,' notes Hannerz, 'is that where participants find satisfying understandings of the world and support for some reasonably high degree of self-esteem' (Hannerz 1969: 111). A rap is a form of 'mythmaking', because 'on myths men ground their beliefs about what moves them and their world' (Hannerz 1969: 115). Although, as we will see below, rap lyrics are inscribed with the style and personality of individual writers, they unmistakably carry a group signature. With their raps, young men create a forum for the exchange and bearing out of a common orientation.

One of the starting points of a rapped myth is a distaste of the immediate living environment. Sarcelles does not seem to be very well suited to young people. Many young people see Sarcelles as a boring place. They have the feeling of being shut in and isolated. Youths often complained about the lack of entertainment facilities in the city, such as discos and cinemas. 'Amusement is impossible to find here,' said one boy about the lack of entertainment in the city. 'Here there aren't any weekends.' 'Life is hardcore, friend,' as Thiburce raps to his generation.

I wander day and night through concrete walls
In bedroom towns, as the papers call 'm
I drag, drag along, with no purpose, no hope
Frustration and desperation are my only companions

Entre les murs de beton je rouille matin et soir
Dans ces lieux que les journeaux appellont cites dortoirs
Je trainè, je trainè ma haine, sans buts, sans espoirs
Avec pour seul compagnon, la frustration, le désespoir

According to Thiburce the *banlieue* is 'a true garbage can, even more ridiculous than a Chernobyl'. 'Tchernobyl' here has a double meaning. On the one hand, it evokes an apocalyptic image of the *banlieue* through the analogy of a nuclear power station disaster. On the other hand, the term refers to a cigarette or joint. In another rap song he describes the hallucinating effect of 'the smell of shit' that pervades the city. He fancies himself in 'a true jungle', full of 'destroyed elevators and pillar-boxes'. Thiburce here uses the word *fonzdés*, a form of *verlan* derived from *defoncé*, which means acting stupid or being under the influence of drugs. These kinds of descriptions repeatedly return in rap lyrics. Although they take daily life as a point of departure, raps are not realistic accounts. They are, to paraphrase Hannerz, 'myths', stylized images of daily existence, in which the city is used as the decor of their exclusion. While rapping with companions in distress, the experience of marginality is turned into a collective product.

Figure 1a Rap text as social script, Thiburce.

Fuck the *dawa* and the *hala*[2]
What are our choices?
To walk the straight and narrow? Or to turn into Pétain?[3]
Only bullshitters make it onto T.F.1[4]
Brothers ain't got no choice[5]
Beeper on their belts, cell phone in their car
The beep beeps sound like a tam tam
Working the phone network
Going all out for the money[6]
Nothing to lose, left alone to hope for
They take it all, leave nothing behind
That's how it ends
My hometown is going down the drain
This ain't no movie, it's reality
This fuckin' system is messing me up, cutting me up
Pay attention, I'm cheatin'[7], not breakin' no balls[8]
Fuck your variety shows, have it your way!
I rap with conviction
Life has to be more than this
But anyway, it's the life I'm living

Foutre le dawa niquer la hala
Qu'est ce qu'on nous laisse comme choix
Choisi le droit chemin? Vénérer Pétain?
Putain d'baratin autant mater T.F.1
Ya pas trop d'choix pour les frères
Beeper à la ceinture, portables dans les voitures

---

[2] 'Dawa' means 'big mess'. 'The city is garbage can', Thiburce explains 'The mess is everywhere'. 'Hala' means 'very beautiful'. Both words derive from Arabic.

[3] Pétain was a French general who collaborated with the Germans under the Vichy government in World War Two.

[4] T.F.1 is a public television channel in France. Rappers see television, along with other media such as newspapers and magazines, as an accomplice of the system; they obscure facts and sell nonsense.

[5] Rappers use the word 'frère' as a form of address to male friends or 'blood brothers'. The word derives from the Anglo-Saxon 'brothers', a designation of members of the same social or ethnic group.

[6] In the original text 'argent' ('money') is turned around into 'gen-ar'.

[7] 'Cheating' is a translation of the popular argot 'magouiller', often used in rap lyrics. The word hints at the preoccupation to fool representatives of the system (see below).

[8] 'Balls' is translated from the argot 'couille'.

Et ça fait bip bip, c'est la danse des Tam Tam
Sur le réseau SFR les frères s'font du gen-ar
Rien à perdre, plus rien à espérer
Et quitte à tout prendre autant ne rien laisser
Ma cité va craquer, pas un film, réalité
Le système m'a magouille viens pas m'casser les couilles
Fuck la variet, j'deal mon rap en barettes
C'est pas ça la vie?
En tout cas c'est celle que j'vis

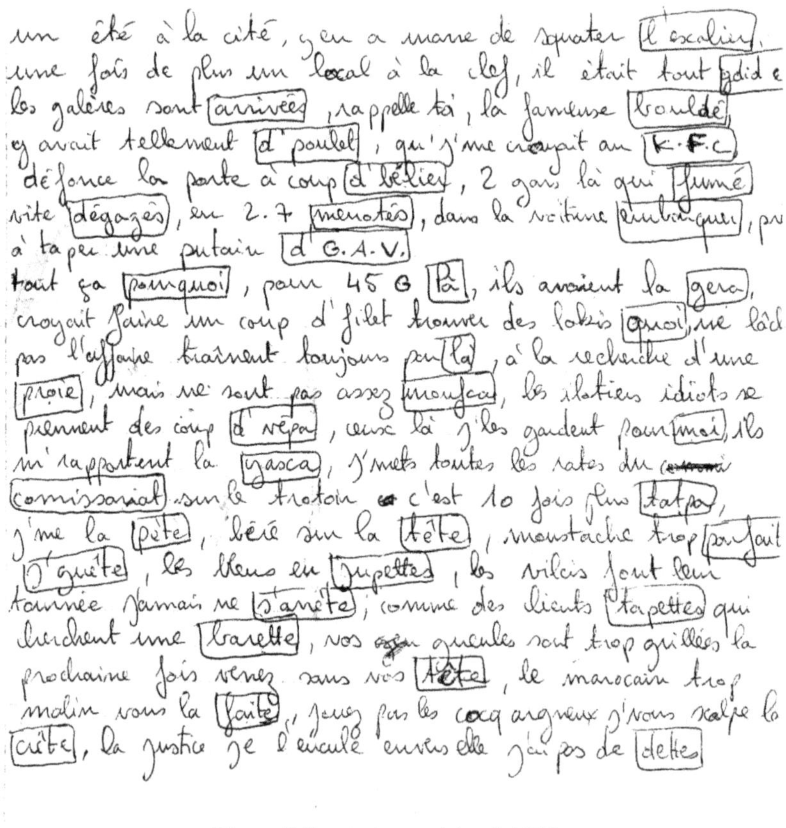
un été à la cité, y en a marre de squater l'escalier,
une fois de plus un local à la clef, il était tout
les galères sont arrivées, rappelle toi, la fameuse boulde,
y avait tellement d'poulet, qu'j'me croyait au K.F.C
défonce la porte à coup d'bélier, 2 gars là qui fumé
vite dégagés, en 2.7 menotés, dans la voiture embarquer, pr
à taper une putain d'G.A.V.
tout ça pourquoi, pour 45 G Pa, ils avaient la gera,
croyait faire un coup d'filet trouver des lokis quoi, ne lâch
pas l'affaire traînent toujours par là, à la recherche d'une
proie, mais ne sont pas assez moufca, les ilotiers idiots se
prennent des coup d'répa, ceux là j'les gardent pour moi, ils
m'rapportent la yasca, j'mets toutes les rates du
comissariat sur le trotoir c'est 10 fois plus tatpa,
j'me la pète, béré sur la tête, moustache trop parfait
j'guète, les bleus en jupettes, les vilcis font leur
tournée jamais ne s'arrête, comme des lieuts tapettes qui
cherchent une barette, nos gueules sont trop grillées la
prochaine fois venez sans vos tête, le marocain trop
malin vous la faite, jouez pas les cocq argneux j'vous scalpe la
cité, la justice je l'encule envers elle j'ai pas de dettes

Figure 1b Rap text as social script, Filipe.

In the above text the rapper reflects on his life and that of his peers. Thiburce states his hometown is 'going down the drain' ('la cité va craquer'), a popular expression for the degradation of the city or neighbourhood. 'Look at us,' he reminds us. 'This is the life we're leading.' In this reflection lies a sentiment which is a breeding ground for resistance. The first-person narrator tries to resist his fate and to struggle out of his daily circumstances. 'The HLM chains me, but I won't let them lapidate me,' Thiburce writes. 'I face the system.' HLM (Habitations à Loyer Modéré) here is used as a designation of the concrete walls of social housing complexes, representing the impenetrable world of the *banlieue*. Here the rapper explicitly addresses his social group and makes an appeal. Central to the above rap text is the notion of *hardcore*. The word derives from the lexicon of North American hip-hop culture and refers to a musical genre, in which fast, aggressive rhythms predominate. However, as used by French rappers, the notion of hardcore refers not to a musical form or style – raps are mainly put to quiet rhythms and sweet melodies – but to the contents of a lyric. Hardcore means 'experienced' and non-commercial. 'Hardcore rap' in this sense stands in sharp contrast to other genres within rap such as the more playful 'rap cool' and the folkloristic 'cultural rap' (Bensignor 1994). In the above rap lyric mention is made of the 'variety shows', a designation for French chansons. 'The theatre, you know, lalalalala,' Thiburce says. For him these cultural forms represent a dated version of French society. In his use of 'hardcore' then lies a social comment. 'When they do not offer it to us,' Thiburce says about his fate and that of his social group, 'then we'll do it ourselves. Justice, police, politics; we fight them. That's it. We are hardcore.' The social rule, hidden in the *rap en barrettes* of rappers like Thiburce, says that when a rapper hands him or herself over to the system, he or she renounces the original intention of the medium. *Ca se passe comme as*, Thiburce writes:

Accompanied by the MC of a thousand and one nights
Deblé Men are never bored like Moda and Dan
I dream of fucking Ophélie Winter too
Penetrating the system like a modern gangster
Dressed in croc skin and chillin' in the Loco[9]
Rappin' futile rhymes that rhyme with tricks
Performing underground and appearing on NRS
Signing on to a major label, dreaming of Vanessa Paradis[10]
Flipping channels on TV, skip Nagui and tell him how ugly he really is

[9] 'Croc' is short for crocodile leather, the material of which trendy belts and boots are made off. The Loco is a Parisian discotheque.

[10] A 'major' is a designation for a large record company.

That's all behind me
It's not so easy, it's all about style
Like my blood brother Bakayoko, MC na ko béta yo[11]
We talk rap yéyé[12], others rap on water
We just go 'lalala' to make it onto fun radio

Accompagné du MC des mille et une nuits
Deblé Men contrairment à Moda et Dan jamais ne s'ennuit
Moi aussi je rêve de me faire Ophélie Winter
Entrer dans le système tout comme un gangster moderne
Me saper en croco et trainer à la Loco
Raper dese rimes futiles qui rimes avec combines
S' la jouer underground et passer sur NRS
Signer sur une major en rêvant de Paradis (Vanessa)
Passer à la télé voir Nagui, lui dire t'es laid
Ça fait partie d'mon passé, des thuns j'e ai pas asser
Tout n'est pas si facile, tout ne tient qu'au style
Le mien est BAKAYOKO, MC NA KO BÉTA YO
On parle de rap yéyé, certains font du rap a l'eau
Nous on fait des la la la pour passer sur fun radio

In the lyric the rapper recalls his earlier preoccupations. The title of the song hints at *Ça se passe comme ça* (*ça* is replaced by *as*), a free magazine, available at MacDonald's and delivering the trendy films, sports, books, videogames and music styles of the fast food empire. Notwithstanding the attraction of the make-believe world of crocodile leather, embodied by popular artists like Ophélie Winter and Vanessa Paradis, Thiburce expresses what distinguishes him as a true rapper: style and originality. Rappers often cite instances when they are finished with an artist. Next to 'senseless' commercial raps, the metamorphosis of Senegalese rapper MC Solaar, a pseudonym for Claude M'Barali, is a startling case in point. On his first records Solaar, who grew up in the southern *banlieue* of Paris, presented himself as a warrior, dressed in camouflaging trousers, his hands put desperately around his neck. Years later, when he turned millionaire, he changed this image for that of a popular artist, poet and intellectual. 'MC Solaar is serene,' Thiburce sanctions. 'These guys are part of the system,

---

[11] Bakayoko is an African football player. 'Na ko béta yo' means 'scoring' and 'making an impression'.
[12] 'Yéyé' is an American style of rapping.

they are no hardcore. You know, they say things like 'respect yourself' (imitating a jolly singing voice). What do they talk about? These guys talk shit.'

Solidarity then functions as a credo, with which a rapper approaches a larger audience. As they reach a bigger audience, however, the contents of a rap lyric shifts. When a rapper comes forward with his lyrics he focuses the outlines of a communal front. Seen in this light, rapping is not an everyday form of conversation. Although it contains many elements of a verbal exchange, rap essentially is a ritualized idiom. Given the appearance of improvisation, the act of rapping is fixed. Rapping is a performance in which the speaker displays his or her vocal and verbal capacities, personality and physical appearance – in short, his or her style. Next to this 'expressive' function, rap also has a 'directive' function (Kochman 1972: 246). Instead of speaking with others, the rapper especially speaks to others. Rap provides him with an instrument for control and manipulation. In the words of Miliani, young men use a 'prescriptive/descriptive discourse' (Miliani 1995: 30). By way of condemnation, accusation, moralizing and provocation of the outside world young men make sense of their world. They defend themselves against a world which robs them of a better fate and prescribes an 'ethos of solidarity'. 'Mecs, those are my friends, people like myself,' Thiburce explains. In this sense, hip-hop has a strong motive. In a rap lyric personal experiences and observations become part of a collective body. 'Within an insecure and unstable environment,' Welz writes, 'a rap session provides an opportunity to construct a shared meaning system, in which events and experiences are interpreted and controlled' (Welz 1991: 325). In the first instance hip-hop raps across cultural differences to cast an anchor in an impenetrable environment (Chobeaux 1994).

The ethos of solidarity links to the position from which a rapper writes and speaks. In the above fragments we could see how Thiburce comments on the conduct of his generation. This perspective is, for example, represented in the logo of his rap group (Figure 2). The drawing is split in two. Above ground, the skyline of a city is visible, illuminated by daylight. Below, the high-rise apartments are reflected in the water, like a city by night. Light and shadow, day and night, these contrasts insinuate the object of his raps. They are written as mirror images, in which the dark sides of urban society are illuminated. In this sense, Thiburce fulfils the role of a cultural broker, someone who writes in the name of others and forms an intermediary between a small social group and the wider society. The rapper wants to approach a large audience, 'to spread his engaged lyrics on a national scale'. Thiburce is in a position to take this stance. He is older than many rappers and has a relatively successful career as a social worker in a community centre in Sarcelles and as the leader of popular rap group.

Figure 2 Mirror images: Logo of French rap group Deblé-Men.

Masking and language coding: transformations and mystifications

Rapping is a symbolic play with words. Within a limited set of stylistic rules – a rap has to fit into a four-four time and have a rhyme scheme – rappers are free to experiment. Rap lyrics, as can be seen in the rap texts quoted in this chapter, are filled with twisted words (*verlan*), made-up terms and expressions (*argot*), and metaphors. Moreover, rappers often use loan words, derived from the Anglo-Saxon language or from the native language of their parents. A rap lyric contains a hybrid language form, combining elements from different languages. These textual disruptions often relate to sensitive cultural information. Next to numerous expressions for police officers, many of them referring to stupid animals (especially ducks and chickens), there exists a whole repertoire of words describing sexual organs, money, clothing, drugs and alcohol. These are significant themes in the social world of young men in the *banlieue*. The language used in rap is therefore difficult to pin down. What is illuminating are the attempts of parents to record the slang of young people into a dictionary (Girard and Kernel 1996). Apart from its educational value, however, these documents are dated as soon as they are published. Although young people also draw from the existing repertoire of *argot* expressions (Colin, Mével and Leclère 1996), words and expressions change at a high speed.

The flexible use of language in rap has a specific function. Rappers build a bridge between two worlds. Their language is transparent for insiders, but functions as an opaque front to outsiders. On the one hand, young men turn to the *communitas* of their own group. However, as they speak to each other, they are also focused on an outside world, a potential public. In their lyrics lies a longing to be heard, to present themselves to a large audience. To enter into the public domain and to build a bridge between private and public domains, a process of distortion is put into practice. The challenge is to infiltrate the power-loaded public domain with a coded language. 'Hitting a straight line with a crooked stick,' as Scott (1990: 153) writes. Voice and text are used as flexible social weapons in the battle with authorities. This is the hidden transcript of rapping. In rap, young men create a discourse which proceeds within the framework of the public transcript of majority speech (the French *langue*) but also within a derivative language, a *parole* mixed with codes.[13] This 'verbal play with majority speech', as it is played out in rap, is related to social borders. A rapper operates within a *speech community*, a universe, marked by language barriers. To quote Gumperz (1973: 219), young men are forming a 'human aggregate characterized by regular and frequent interaction by means of a shared body of verbal signs and set off from similar aggregates by significant differences in language usage'. Within this speech community language functions as a marker of group identity.

---

[13] The words 'langue' and 'parole' derive from the philosophy of Ferdinand de Saussure. With 'langue' he refers to the communal and constant grammatical rules of a language. A 'parole' is a historical conception and refers to a language as it is practised in daily life.

In addition to the message, rappers can also disguise the narrator. When approaching a large audience, the rapper assumes a second identity. For example, Thiburce operates under the name of Bursty or BURTYX – in which his name is spelled from back to front – alias 'the rotten intellectual brain'. To promote their first CD, his rap group issued a press communiqué, in which Thiburce is described as 'the chief of the tribe', 'Majestical, Brave, Suspicious' and as 'The Old Martial Man, respected by his people'. The other members of his group, Deblé-Men, also use pseudonyms. Daoudix, alias Daoud le MC, is called the rapper of 'a thousand and one nights', the 'small war-minded man with an evil spirit and living intellect', who undertakes 'all dangerous missions without hesitation'. O.W.I.X. presents himself as the 'Spiritual Circuit Breaking MC of darkness'. He is called the 'Big Amateur of Verbal Destruction', who 'seeks adventure'. Finally, TRIKPA DJIX, alias 'The Musical Druid', is the deviser of the group's music and concept. As soon as rappers build a bridge between a local circle and a larger audience, a remarkable change occurs. As they disclose themselves and their message, rappers make use of disguises, camouflages and pseudonyms. This phenomenon points us to the tension between hidden and public transcripts. Although bound by social conventions, people within the private domain can speak relatively freely. As soon as they enter a public domain they are under surveillance. The rap group Ministère A.M.E.R., an abbreviation of Association Musique Et Rap, enjoys a reputation with regard to public propaganda. Two lawsuits were issued against the group as a result of the content of two of their songs (Prévos 1998). Next to 'Sacrifice de poulet' (see below) *Brigitte (femme de flic)* was considered to be a blasphemous and offensive song. The story describes the sexual escapades of rapper Stomy Bugsy, who poses under a pseudonym, with a daughter of a commissioner and the wife of a police officer. The group members themselves describe the song as a 'kamasoudrap'. In the background we hear a satisfied and groaning woman, 'yearning for a nigger'.

Striking poses: fantasy and the romanticization of a marginal existence

Rap functions as a format within which men are able to publicly, in the presence of their friends, display their verbal and vocal capacities. In a closed circle, young men perform an internal and mutual competition, in which the pace of speaking, articulation and inventiveness in the choice of words are turned into important criteria for measuring a good rap. With a combination of style and content young men demand respect from their

fellows, which they lack in mainstream society. A dominant theme in the
banlieue rap is the exploitation of a marginal position and a preoccupation
with social status. To claim personality rappers often strike a macho pose.

> Sarcelles criminals are pimpin' the city
> Wake up, take a good look at this place, we're all chillin'
> The cops on their beat, girls lookin' lovely[14]
> Caught a rat and taught him a lesson
> Let me give you my number or else let's go to a motel[15]
> That's where I make angels scream
> I'll show you ecstasy, my sting is harder than a bee's
> My rock is bigger than a Clan Campbell bottle[16]
> I'll let you lick my chest and suck my scars
> The real will become unreal
> I'll show you the seventh heaven
> Or better yet, the 27th floor
> Or all the way to the roof
> But be sure to wear lace and suspenders
> Whether your name is Giselle, Christelle or Estelle
> Whether you're from Sarcelles, Villiers-le-Bel or Chavnontel

> *Les criminals de Sarcelles une fois de + foute le bordel*
> *Ouvre tes oreilles dans ma ville, l'atmosphere et belle*
> *Les keufs a l'appel, les perelles les plus belles*
> *Les rates j'ai ramasse a là pele*
> *J'tépelle mon phonetel ou sinon passé a mon etap hotel*
> *La j'aide ploirai mes ailes*
> *Tu vera mon dare plus puissant que ce lui d'une abeille*
> *Avec un dian etre plus large qu'e celui du bouteille Clan Campebell*
> *J'te lecherai les manelles et après toi tu pompera mon scalpelle*
> *Avec moi que du reel pas d'irreel*
> *J'te ferai monter au 7ème ou plutôt au 27ème ciel*
> *Mais surtout vient habiller en dentelles et en porte jarettelle*
> *Que tu t'appelle Giselle, Christelle ou Estelle*
> *Que tu vienne de Sarcelles, Villiers-le-Bel ou Chavnontel*

---

[14] In the original text 'police' (*keufs*) and 'girls' (*perelles*) are written in argot.

[15] A 'motel' ('Etap Hotel' in the original text) is a cheap hotel, often found at the exit of a motorway.

[16] Clan Campbell is a brand of whisky. The size of the bottle goes without saying.

In the above fragment Filipe sketches a rather 'hot' atmosphere. The first-person narrator seduces a girl, shows off his criminal past and presents to her his dangerous sexual weapons. In the text, genitals are described in veiled terms as 'angels' and 'diamonds'. The storeys of apartment buildings indicate the levels of ecstasy in which the girl finds herself. For that matter, women in rap are often caricatured or used as objects for the display of obscenities, sexual actions and intercourse. Here we see a tendency for the 'oppressed' to oppress others. Making insinuations to the size of a testicle, the potential of a scrotum, the satisfaction of sexual intercourse and the repertoire for making passes are all played out as symbols of virility. Next to displaying the subordinate position of women, rappers express a longing for money and social status. 'We are prisoners of money', as the group Ministère A.M.E.R. said on this theme. A leitmotif of many rap lyrics is the completion of a drug deal. For example, in the following text Filipe describes the way 'he', a fictive character, survives in daily battles. Driving a Ferrari Testarossa at full speed, and equipped with 'kleber-feet', a brand of non-skid tyre, he gathers 'the glory' and 'installs' himself 'in the contest'.

> Just like Pablo but without touching dope[17]
> Earning big bucks, raking in zeros[18]
> I don't want to end up poor
> Tasting affluence is a Moroccan ambition

> *Comme Pablo, sans cheveu de pedo*
> *Avoire du zeillo, accumuler les zéros*
> *J'ai déjàs goutter à la galère, j'veux pas finir comme un vrepau*
> *Tater à la richesse est l'ambition du maroco*

The above lyric describes the satisfaction of making a quick buck during a *biz de dyname*.[19] Filipe evokes a character enriching himself at the expense of others. 'Mediocre people', as he calls them, bite the dust for him 'like a mouse in a cage'. Filipe employs a much used trick, by which he makes himself triumph over others. However, there is no royal road to status. As we

---

[17] 'Pablo' is the front name of drug baron Pablo Escobar. 'Dope' is a designation for drugs and is derived from 'doper'. In the text the word is spelled as 'pedo'.
[18] 'Earning big bucks' is a translation of 'avoire du zeillo'. 'Zeillo' is 'ozeil' turned around. The 'zeros' refer to low life persons and to the zeros in a large sum of money.
[19] 'Biz' is an abbreviation of 'bizness' or business. 'Dyname' refers to 'dynamite', which is argot for 'cocaine'.

can read in the next fragment, living in the urban quarters of the *banlieue* is like going through the school of hard knocks.

Till now I was a quiet high school student
Didn't try messing with the easy money
I didn't worry about the days passing by
But just wait, brother
One day I'll dive into the bizillicite[20]
Don't care about climbing fast
'Cause nowadays jealousy is free in our 'hood
Don't envy those friendships breaking up over shit
A few small tokes on the mythic bebar bleuf
I fear the guita,[21] don't have time for chicks
I make my own party, invite only the cops
Who want me to doze off in this new town
The kid I once was, has grown up
These days I talk only in kilos,[22] no discretion
The police see me as the godfather of the cité[23]

Jusque là j'étais un callègieu tranquille
Peusais pas à l'argent facil
Me souciais pas des journée qui défilent,
Aimi soit-il,
Or un jour j'me lance dans le bizillicite
Pas l'attention de montre vite,
Car de nos jours dans nos quartiers la jalousie est gratuite
Pas envie de voir des amitées définites pour du shit
Quelques petit coups de bebar bleuf
J'peure trop à la guita, pas de temps pour les meufs
J'fais même pas la teuf,
J'évite juste les keufs qui veulent me faire dormer dans un endroît tout neuf
Le petit qu' j'étais à bien grandis

---

[20] In 'bizillicite' (meaning 'illicit business') 'bizness' and 'illicit' are put together.

[21] Guita is a Portuguese monetary unit.

[22] 'Talking in kilos' is a translation of 'parler en loki', which means 'dealing drugs', but also connotes the practice of addressing people in a low key manner. 'Loki' is verlan for 'kilo'.

[23] 'The godfather of the city' is a translation of 'le parrain de la técis'. 'Parrain' is argot for a 'person of no importance'. In 'técis' 'cité' is turned around.

Aujourd'hui j'parle en loki,
Tait en restant scrèdi meme si
Les geuts me vois comme le parrain de la técis

Filipe describes an imaginary process of social success. Boasting and showing off are the idiom of rap. In their lyrics rappers create a substitute for an impenetrable world. They display themselves as tough and streetwise characters, holding out in a stubborn environment. 'I', as a rapper, wearing a monk's cowl, presents himself in *Sacrifice de Poulets* (Sacrifice of a Policeman):

I display all the symptoms of a bad ethnic
Antipathetic, sadist, allergic to police
Even in a crowd I wear a monk's hood
The youth listen to me
In the school of the street I am a pro

Moi j'ai toutes les caracteristiques du mauvais ethnique
Antipathique sadique allergique aux flics
Meme dans la foule je porte la cagoule
Les plus jeunes m'ecoutent
Dans l'ecole de la rue je suis un prof

The narrator in the above rap text is a recurrent character in rap lyrics. Here is a rapper who sets himself up as the leader of the troops on the field of battle. 'Once again, one time to many,' he tells his listeners in *Au dessus des lois* (Above the law), 'the voice of the nigger addresses the problem, the dilemma against the system ... I say no to the police, the army, the street ... and declare myself above the law ... I go against the current ... [because] above the law I don't believe I will be the prey they are sending to the front, a sort of charcoal to burn, grill, end like a medal, mutilated to march the Champs. No.' Champs refers to the Champs Elysées, pointing to the mental distance between suburb and inner city. These lines indicate a fourth recurrent principle in rap: the longing for a symbolic revenge on those in authority.

Turning up worlds

Rap carries an attitude to life. 'An eye for an eye, tooth for a tooth', as they call it, 'the law of retaliation'. In their lyrics young men search for a confrontation with authorities. They compare their position with those of other marginal or repressed groups and make an analogy with wars and battles that are similar to those fought in the *banlieue*. In descriptions of their own 'fucking neighbourhood' they refer to the situation in North American ghettos, the occupied territories in the western bank of the Jordan, the apartheid regime in South Africa, the Occident and the slave trade, World War II, the nuclear threat and the bomb on Hiroshima, and to Ali Baba and the gate of Sesame. This identification with and taking warnings from other marginal or oppressed groups is a much used device in rap. Elflein, who found a similar phenomenon in Turkish rap in Germany, describes rap as part of a 'sub-cultural discourse of dissidence'. 'Dissidence,' he writes, 'means the construction of a worldwide community of all those who are deprived of power – regardless of their social, political and economic differences. In particular, such dissidence takes no account of whether exclusion is a symbolic act of free will (which can be reversed at any time) or a compulsory reality of life' (Elflein 1998: 258; see also Bennett 1999; Maxwell 1997). With their identification with victims of a restrictive environment, however, young men will not describe themselves as losers. In the 'jungle' or 'prison', to use two other analogies, they make up their own rules. 'In a world of idiots, who defend our laws. In a world of idiots the police is outlawed,' as they claim.[24] See the following statement in which a rapper heads for a battle with the police.

> The police and the media are out for prestige
> But the world is mine, my name is Tony Montana
> Man, these times are mystical
> Before I let my murderous pulses flow
> I'm addressing my last prayers to everyone
> Friend, ask God to let you live
> I'm praying to the devil to let blood flow
>
> It's three p.m.
> The place has cleared out, civilians scattered
> Who's David? Who's Goliath?
> The journalists and their cameras on the ground
> I swear on my mother's life, the guerrilla is degenerating
> Sirens, smoke bombs, spreading sulphur
> More poison, we're stuck in a vortex
> The crowd has dispersed, no one wants to leave

---

[24] 'Idiots' is a translation of 'bâtard', in the original text spelled as 'tarbas'.

They want paradise, but don't want to die
It's too early to party, too late to retreat
It's full moon tonight and my luck is good
And all the punches I pulled hurt the natives
Tonight I'm healthy, I sacrificed a 'chicken'

Les keufs veulent le prestige
Comme le font les medias le peuple suit tous mes pas
Le monde est a moi, je suis Tony Montana
Mec le temps est mystique
Avant de laisser faire mes pulsions meurtrieres
J'adresse au tout puissant mes dernieres prieres
Mec demande a Dieu de rester vivant
J'appelle le Diable pour faire couler le sang

3 heures du mat partout ca calte les civils s'eclatent
Qui est David? Qui est Goliath?
Des journalists a terre des cameras par terre
Sur la vie de ma mere la guerilla degenere
Sirenes fumigenes ca sent le soufre
Plus d'oxygene nous sommes dans le gouffre
La foule se disperse personne ne veut partir
Ils veulent le paradis mais ne veulent pas mourrir
Trop tot pour festoyer trop tard pour reculer
Ce soir la lune est pleine ce soir je suis en veine
Et tous les coups que j'assene font mal a l'indigene
Ce soir j'ai la sante je vais sacrifier un poulet

Remarkable in the above lyric is the use of metaphors, including biblical, instead of *argot* or *verlan*. The rapper comes forward, addressing his 'prayer' 'to everyone'. Illustrative for the use of metaphors to mask information is a rap, written by Thiburce, as printed in an issue of a regional magazine. In *Le Fléau* (The Scourge) he denounces the French political system. However, this time the lyric contains no slang, but is filled with hyperboles, aphorisms and euphemisms. The French Republic is called a 'bastion of hate', working 'under the aureole of the patriot myth' and with the 'flavour of arsenic'. 'Stand up, come on, stand up!' he addresses the inhabitants of the *banlieue*. 'Everyone, join the dance.' Although in real life they are reserved an inferior position, in their raps young men seek an opportunity for revenge. There are many expressions being used in rap which refer to the moment for striking. For example, rappers refer to an 'arrival at the frontline', 'coming out', 'launching themselves', 'carrying a mission', 'taking a stance', 'pointing curtain fire' or 'preparing a cigar'. The target they have in mind is called 'the system' in all its appearances. In addition to journalists and police officers, different elements are ridiculed in rap lyrics, such as the justice system, politicians, television companies and architectural objects. They are the exponents of a larger, impenetrable whole. Unlike in real life, these personifications will taste defeat.

The guerrilla, as described above, forms a symbolic battleground. Against 'the cowards with their guns and rifles', the rapper brings in his verbal weapons. These 'murderous pulsions' do not refer to physical violence, but to a torrent of words. As Thiburce writes in a comparison between the *banlieue* and Vietnam, fired and gassed by napalm: 'The burr of their shovels, the police excels. We fight their charges by our own means.' Next to the veiled retaliatory action, the resistance against authorities finds another subtle translation in rap. Many rap lyrics are grafted onto the inventiveness and originality by which men turn around the established order. *Faire le truc*, as a common expression goes. The main objective is to undermine the authorities as a *trickster*, to play a nasty trick and outsmart the other. Rap lyrics are filled with words that refer to this mission. Terms and expressions such as 'truc', 'cheating', 'a curved road', 'unravelling yourself', 'stymie someone' and 'putting things straight' all point to the textual practice of making a fool of representatives of the system, mislead them and get round their 'ambushes'. In the following text we see how this process works out.

Summer in the city
We stop invading elevators
Once again the centre is locked, it's brand new
The galères have come, remember when we met up?
All those chickens, I swear on Kentucky Fried Chicken
The front door slams, two guys smoke quickly without exhaling
They're handcuffed and carried off, ready to beat up one of those garde a
vue fools
For what? 45 grams of shit, man
They thrive on their anger, planning to make an easy catch, find some
kilos
Born weak, no fight in them, that's how it drags on
Looking for prey, but the camouflage failed
The riot police took some of our shit

Them, I let 'em look out for me, they bring me money
I've met all the rats in the commissioner's office
On the sidewalk, that's ten times as stupid
I'm burning the shit, it's blowing my mind
This moustache looks good on me, I'm checking out the girls' skirts
The civils walk their beat, they never stop
Just like those crazy customers looking for a beret
Next time I'll burn their faces, they cackle
The Moroccan is starting to boil over, won't play cocq argnaux no more
I'll scalp your cockscomb for you
Justice, I'm cheating it
Money, I don't have it

Un été à la cité
Y en a marre de squatter l'escalier
Une fois de plus un local à la clef, Il était tout gdid et
Les galères sont arrives, rappelled toi, la fameuse bouldé
Y avait tellement d'poulet, qu' j'me croyait au K.F.C.
Défonce la parte à coup d'bélier, 2 gars là qui fume vite dégagés
En 2.7 menotés, dans la voiture embarquer, prés à taper une putain d'G.A.V.
Tout ça pourquoi, pour 45 G là
Ils avoient la gera, croyait faire un coup d'filet, trouver des lokis
Né lâche, pas l'affaire traînent toujours par là
À la recherché d'une proie, mais ne sont pas assez moufca
Les ilotiers idiots se prennent des coup d'vépa
Ceux là j'les gardent pour moi, ils m'rapportent la yasca
J'mets toutes les rates du commissariat
Sur le trotoir, c'est 10 fois plus tatpa
J'me la pète, sur la tête
Moustache trop parfaite, j'guête les blues en jupettes
Les vilcis font leur tournée, jamais ne s'arrète
Comme des clients tapettes qui cherchent une barette
Prochaine fois venez sans vos tête
Le Marocain trop malin vous la faite, jouez pas les cocq argneux
J'vous scalpe la crête
La justice je l'eucule envers elle
J'ai pas de deltes

This rap sketches a daily routine. Filipe uses the word '*galère*', a name often used as a designation of a group of people hanging round. Some friends stroll through the city, looking for some fun. Influenced by dope, the men praise themselves. They recall moments when they conquered and fooled the police 'fast and without puffing smoke'. The 'civils' refer to the *garde civile*, who patrol the streets. As many times before in the 'enduring fight' they come off worst. The corrupt 'chickens' and 'rats' got their number. The text once more is filled with twisted words. 'Pavé' (shit, drugs) is spelled as 'vépa', 'cayas' (money) as 'yasca', and 'tatpa' turned around ('patat') means 'fool'. However, the resentment that lies in this reconstructed story functions as an overture for future action. The rapper bends things to his will. He writes a script in which he summons himself and others to start a retaliatory attack. The next time, the rapper states, he will 'burn their faces' and 'scalp the cockscomb'. Like a Kentucky Fried Chicken, the police officer is ready to be served. When the biter is bit revenge will be sweet.

Conclusion

Rapping is charged with symbolism. This paper shifted attention from the materialization of symbols in the physical environment to the less tangible symbolism of a social and textual practice. Moreover, the analysis of rapping, as practised by young people in the *banlieue*, showed the contested nature of urban symbolism. Symbols simultaneously unite and divide. We saw how youths make use of the hidden transcript of a sign language in which they infiltrate the public domain with their messages in a concealed manner. Rapping is a ritualized form of sociability. Rap texts form a sort of folk utopia, stylized stories of daily reality. Where the immediate environment leaves little to the imagination, youths take refuge in fantasized worlds. With their identification with the ghetto and with other marginal groupings, the symbolic revenge on authorities and the implementation of an internal and mutual competition, young people bring their own discourse to life. In this distortion of reality a remarkable reversal process takes place in which a marginal position is elevated to a status object. Youths use their masculinity, money, status, sexuality, and illegality as a protest to claim their dominance. They bring to life a substitute world within which they describe themselves as victors. Rapping thus shifts attention to the contested nature of a symbolic order. The coded French rap language forms an ambiguous discourse in which youths appear in public to demand the respect and build up the self-confidence that ordinary society denies them. Youths demand their place in the urban system, but at the same time they position themselves outside it. Here lies the quintessence of rap, reaching beyond the boundaries of the local community into a global diaspora of marginal groups as well as being practised as a local form of expression, anchored in a weekend-less environment.

References

Abrahams, R.D. (1974) 'Black talking on the streets.' In: R. Bauman and J. Sherzer (eds), *Explorations in the Ethnography of Speaking*, pp. 240–62. New York: Cambridge University Press.

Bennet, A. (1999) 'Hip hop am Main: The localization of rap music and hip hop culture.' *Media, Culture and Society*, vol. 21, pp. 77–91.

Bensignor, F. (1994) 'Le rap français vers l'âge adulte.' *Hommes et Migrations*, vol. 1175, pp. 52–6.

Blair, M.E. (1993) 'Commercialization of the rap music youth subculture.' *Journal of Popular Culture*, vol. 3, pp. 21–33.

Chobeaux, F. (1994) 'L'identité collective de jeunes en difficulté d'insertion sociale.' *Hommes et Migrations*, vol. 1180, pp. 23–9.

Colin, J-P., J-P. Mével and C. Leclère (1996) *Dictionnaire de l'Argot*. Paris: Larousse.

Dimitriadis, G. (1996) 'Hip hop: From live performance to mediated narrative.' *Popular Music*, vol. 15, no.2, pp. 179–94.

Elflein, D. (1998) 'From Krauts with attitudes to Turks with attitudes: Some aspects of hip-hop history in Germany.' *Popular Music*, vol. 3, pp. 255–65.

Geertz, C. (1973) *The Interpretation of Cultures: Selected Essays*. New York: Basic Books.

Girard, E. and B. Kernel (1996) *Le Vrai Langage des Jeunes Expliqué aux Parents qui n'entravent plus rien*. Parijs: Albin Michel.

Gumperz, J. (1973) 'The speech community.' In: P.P. Giglioli (ed.), *Language and Social Context: Selected Readings*, pp. 219–31. Harmondsworth: Penguin.

Hannerz, U. (1969) *Soulside: Inquiries into Ghetto Culture and Community*. New York: Columbia University Press.

Kochman, T. (1972) 'Toward an ethnography of black American speech behaviour.' In: T. Kochman (ed.), *Rappin' and Stylin' out: Communication in Urban Black America*, pp. 241–64. Chicago: University of Illinois Press.

Maxwell, I. (1997) 'Hip hop aesthetics and the will to culture.' *The Australian Journal of Anthropology*, vol. 1, pp. 50–70.

Miliani, H. (1995) 'Banlieues entre rap et raï.' *Hommes et Migrations*, vol. 1191, pp. 24–30.

Mitchell, T. (1998) 'Australian hip hop as a "glocal" subculture.' Paper presented at the Ultimo Series Seminar, UTS, 18 March.

Nas, P.J.M. (1993) 'Introduction.' In: P.J.M. Nas (ed.), *Urban Symbolism*, pp. 1–12. Leiden, New York and Keulen: E.J. Brill.

Prévos, A.J.M. (1998) 'Hip-hop, rap, and repression in France and in the United States.' *Popular Music and Society*, vol. 22, no. 2, pp. 67–84.

Scott, J.C. (1985) *Weapons of the Weak: Everyday Forms of Peasant Resistance*. New Haven and London: Yale University Press.

—— (1990) *Domination and the Arts of Resistance: Hidden Transcripts*. New Haven: Yale University Press.

Toop, D. (1984) *The Rap Attack: African Jive to New York Hip Hop*. London: Pluto Press.

Turner, V. (1982) *The Ritual Process: Structure and Anti-structure*. Ithaca, New York: Cornell University Press.

Wacquant, L.J.D. (1993) 'Urban outcasts: Stigma and division in the black American ghetto and the French urban periphery.' *International Journal of Urban and Regional research*, vol. 3, pp. 366–83.

Welz, G. (1991) *Streetlife: Alltag in einem New Yorker Slum.* Frankfurt: Notizen.

Discography

Deblé-men (1996) *Les Rates, les Wacks, les Bacs.*

Various artists (1996) *L'Alliance du Nord: R.E.P.R.E.S.E.N.T.* L.G.O. Production ADN001.

MC Solaar (1994) *Prose Combat.* Polydor 523979-2.

Ministère A.M.E.R. (1992) *Pourquoi tant de Haine.* Musidisc MAM Production 109172.

Ministère A.M.E.R. (1994) *95200.* Musidisc.

Nég'Marrons (1997) *Rue Case Négres.* Sony S.M.ALL. 487256-2.

Chapter 9

SELLING THE CITY
Representations of a transitional urban community

Roelf ter Berg

Introduction[1]

*To put it polemically, there is no such thing as a city ... I would argue
that the city constitutes an imagined environment. What is involved in
that imagining – the discourses, symbols, metaphors and fantasies
through which we ascribe meaning to the modern experience of urban
living – is as important a topic for the social sciences as the material
determinants of the physical environment.*

(James Donald, 'Metropolis: The City as Text', 1992)

In 1994,[2] the City of Cape Town, like the rest of the country, stood at the
beginning of a transformation. It found itself again in the warm, but
somewhat tentative, embrace of the international community. And, having
broken the shackles of apartheid, a new but unknown future lay ahead, for

---

[1] This article is an edited version of part of my unpublished MA thesis *The City as Text:
Images, Space and Symbols in Cape Town, South Africa*. Written in 2004 under supervision of
Prof Dr P.J.M. Nas, this thesis combined insights from the fields of 'Urban Symbolism' (see
Nas 1984, 1993), 'The City as Text' and 'Global Cities' with writings on the post-industrial
transitional city and a preliminary study by Margot N. te Velde and was largely based upon an
extended period of fieldwork in Cape Town from November 2002 to April 2003. The focus of
this study, due to unsafe fieldwork conditions, was upon Cape Town's central city,
representations of the city and metropolitan spatial patterns. An in-depth study of symbolism
in the periphery unfortunately was not possible in the circumstances.
[2] In 1994, South Africa's first democratic elections were held.

many problems still faced the city, not the least of these the rebuilding of its reputation. In order to attract the investments necessary to further the stagnant economy and address the pressing socio-economic issues, it was of prime importance to regain the trust of the world. In line with international developments, the City of Cape Town thus brought to life several institutions that had the explicit task of producing a coherent and attractive image to be used mainly for external consumption.[3]

As one can observe cities all over the world embarking on conscious strategies to promote themselves with the aim of luring investors, tourists, businesses, industry, events and international organizations of power and prestige, Cape Town thus set out on a similar course. It tapped into an international discourse on the reconstruction or rewriting of the city. For, as signs and symbols have always been fundamental to social life, this seems to be even more so for the leading cities in today's globalizing world. In view of this, some authors have even been led to reinvoke Kodak's 1990s campaign slogan: 'image is everything!' (Short 1996).

Therefore, in order to realize its full potential, Cape Town found that it had to present the world and the nation with a new face. Dismissing with the negative connotations of the apartheid state and the sad prospects the international pariah status carried with it, the city faced the challenge of rebuilding its image.

Taking my cue from the works of Lynch, Short, Zukin and Nas amongst others, this article is concerned with the construction of this new image and the elements that constitute it. Thus, this case study of the City of Cape Town illustrates the indivisibly entwined nature of imagined environments and the material determinants of public life. True to its title, it is indeed about 'selling the city'. What are the key elements constituting the city's new identity? Who is spreading the message? And how do the city's promotional efforts relate to meta-narratives and super images?

Logos

First of all there is the process of name-giving. For the name of a city often reflects its historical roots, a unique characteristic, its ambition, hopes, ideals or a combination of these. The naming of a city thus is an important event, often commemorated in myth and ceremony (Nas 1984). On the other hand, names given to a city can reflect outsider perspectives and changing imageries. The City of Cape Town, for that matter, is no different and throughout its history a variety of names and adjectives reflected the many

---

[3] The question of internal representation will be dealt with later.

faces of this diverse and cosmopolitan city. The 'Fairest Cape', 'Cape Town', 'Cape of Storms', 'Tavern of the Seas', the 'Mothercity' and 'Cape of Good Hope' for example. But mocking the easygoing attitude of the city's inhabitants, it is also called 'Slaapstad'.[4] Thus, the city's many names refer to geographical location, culture and history, amongst others. Recently, it was the city administration's choice to be named 'The Unicity' that is very telling, though, and reflects its preoccupation with the social theme of unity.[5]

Of even greater importance, though, are the dynamics of the city logo. For it is through official symbols such as the city's seal and logo that powerful images can be conveyed in, or rather because of, a simple format.[6] Figures 1 to 5, for example, show a range of official city logos, public recognition of which is fairly high, as they figure or have figured prominently in all kinds of city campaigns and are frequently used in diverse media.[7]

Not surprisingly,[8] the nature theme especially seems to be of great importance and, for a large part, it dominates external and internal representations of the Cape character. But the other themes mentioned above are easily identifiable as well. Even the link with the rest of the nation is established in the figuring of the Protea,[9] the national flower, in the logo of the Cape Town tourism bureau.

Being the City of Cape Town's official logo,[10] Figure 1 especially makes its appearance all over the city. It is on the cover or letterhead of official city publications and pronouncements; it has high visibility as an emblem on flags and posters all over the metropolis; and it figures prominently on the

---

[4] Id est, Sleep City.

[5] The Unicity is an amalgamation of the former independent municipalities of Blaauwberg, City of Cape Town, City of Tygerberg, Helderberg, Oostenberg and South Peninsula.

[6] The power of logos is easily illustrated by the ease with which we identify organizations by their symbols and the adjectives and associations that are immediately generated in the back of our minds. A logo, no doubt, often is one of the most recognisable and well-used ways of representing or imagining the city: the Big Apple or New York being a prime example.

[7] A logo is one of the most recognizable and well-used ways of representing or imagining the city.

[8] In light of the observations made by te Velde (1998) on the importance of the nature theme.

[9] The Protea is a powerful national symbol with a high visibility. In a natural way the Protea conveys a message of beauty, purity and strength. Thus embraced by many, the Protea has become a powerful symbol of national unity and strength.

[10] As part of the Joint Marketing Initiative (JMI), which I will later refer to in greater detail, a new logo for the city is currently being designed. But it is unclear when this project will be finished. For now this logo, already designed by the Cape Town municipality, thus remains to be used in an official capacity.

vehicles and clothing of city employees. It is by far the city's most widely recognized logo and it essentially shows up just about anywhere. Furthermore, in its simplicity, it is very clear. The colours – green, red, blue and yellow – refer to the national colours and thus connect the city to the land. Meanwhile, the city is symbolized by its most widely recognized icon: Table Mountain and its smaller neighbours, Devil's Peak and Lion's Head. The majesty of this image is highlighted by the overall shape of the logo, which represents a waving flag, thus instilling the image with a sense of pride. Overall, though, this logo remains rather simplistic and presents us with a somewhat narrow frame of reference with an undeniable focus on the nature theme.

The other logos are more elaborate, but all place Table Mountain at the centre stage. In a way, the city's natural skyline dominates external representation as it dominates the city, both from a physical perspective, the sheer mass of Table Mountain rising up behind and towering over the CBD (Central Business District) and historical centre, as well as from a symbolical perspective. Te Velde (1998), for that matter, already commented on the mountain's extraordinary power and importance for the city's inhabitants and its central place in Capetonian identity. And it is interesting to see that, clearly, this power has not lessened in recent times. Table Mountain still remains the city's most recognized icon as well as its favourite image. The central usage of this powerful symbol in the city's iconography, therefore, cannot be a big surprise, especially as one realizes the recognizability[11] of this idiosyncratic, highly characteristic and well-known feature of the city's natural surroundings. The mountain truly is a majestic symbol rooted in nature, but undeniably linked to the City of Cape Town and its history. Thus, the picture of Table Mountain lends to the city an image of all the good associated with nature and at the same time fills the mind with an anticipation of grandeur and majesty. The Table Mountain – or at least its mental representation – undeniably lies at the heart of this city's identity and representation. Images of Cape Town all seem to have been build around this powerful icon as the city was build around the mountain.

Logo 2 and 4 used by the Cape Town Tourism Bureau then elaborate upon the nature theme, and images of the ocean and the sun are incorporated. Thus, they are trying to convey a powerful message of stunning natural surroundings combined with a warm and pleasant climate. In its attempts to lure tourists the Tourism Bureau apparently focuses on Cape Town as a

---

[11] Besides the figure of Nelson Mandela, probably no other South African symbol is as widely known as Table Mountain.

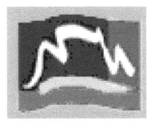

Figure 1 Official logo of the City of Cape Town, 2003.

Figure 2 Logo used by Cape Town Tourism Bureau on 2003 website.

Figure 3 Cape Metropolitan Council (CMC) logo found on website, 2003.

Figure 4 Another logo used by Cape Town Tourism Bureau on their website, 2003.

Figure 5 Logo Partnership in Delivery used by the City Council, 2003.

leisure destination. The incorporation of the Protea in the Cape Town Tourism Bureau logo moreover is a reference to the city's bond with the nation as a whole.

Figure 3, the City of Cape Town administration (CMC) logo, also shows the mountain and the ocean, while the white cloud-like shape – the Table Cloth[12] – represents the wider Cape Metropolitan Area. The shape of the Cape peninsula and False Bay are easily recognized. Designed after the amalgamation of the former independent municipalities, this image clearly relates to the theme of unification in so far as it shows that Cape Town is more than the mountain and the CBD alone.

It is the last picture, though, that is of the greatest interest. Logo 5, designed to accompany the city administration's Listening Campaign,[13] represents both the purpose of the campaign and a shift in administrative thinking. During the course of interviews, several administrators and politicians told me that public office these days is slowly starting to realize that to work for the public no longer suffices. The public has to be informed when and where progress is made and two-way communication is essential. Thus, a realization is dawning that it is not enough to market the city externally. It has to be marketed internally as well. The people included in the new logo, in a very crude fashion depicted by lines and dots at both sides of the picture, partly reflect this change in attitude. On the other hand, the inclusion of people in the logo also results from the recognition of the diversity of the city's inhabitants as a real asset. Commissioned by the City of Cape Town, The Cape Character Study (November 2000) has been very instructive in that matter. Finally, the administration seems to realize that a city is made up as much by its citizens as by its buildings and infrastructure. And they recognize and admit that the population of a city, if not anything else, plays an important part in attracting tourism and investment by 'making people feel at home'. The city is thus understood to be a social as much as a physical entity.

Moreover, something else that differs from earlier logos is the inclusion of a stylized image of the CBD. The logo has become less one-dimensional and the city is presented as just that, 'a city'; no longer a non-defined spot under the clouds of Table Mountain, but a diverse, exciting and invigorating entity on its own, a place in itself worth the journey, but also, rather coincidentally, sited in an enthralling natural environment.

---

[12] In summertime clouds, coming from the South roll over the mountain and slide down its northern slopes, where they disappear. Locally, this phenomenon is called 'the Table Cloth'.

[13] During my stay the local authorities launched an initiative they termed 'Listening Campaign'. The main purpose of this campaign was to start a dialogue with the city's inhabitants.

The new design thus incorporates several facets of the city that were not included in earlier logos and this change both illustrates and reflects a transition. At the centre of power a marked difference in attitude is evident. Recognizing the strengths of its diversity, the city is trying to start development by forming partnerships with business and community organizations. As stated earlier, the city and its citizens have thus been recognized as assets and potential partners in development.

Figure 6, figuring at the bottom of each page in publications relating to the Listening Campaign, is even clearer. Besides the mountain, the CBD and the people, there are also stylized representations of the city's beaches, right, and the Cape Flats, left, in this image. The city thus presents itself as a multifaceted metropolis wherein every citizen counts. An image is presented of a city united in diversity, respectful of its multicultural populace. But unfortunately this is a somewhat romantic idea that does not yet correspond with reality, as socio-economic and racial divides still run deep.[14]

Figure 6 City image: Listening Campaign (Source: Partnership in Delivery, 2003).

Marketing the city: an integrative process

The development of logos demonstrated above already showed that the City of Cape Town has over the past decade been very conscious of its image. Mainly for external consumption at first, an effort was made to build a coherent, consistent and favourable image of the city. This effort has been characterized by growing levels of co-operation amongst public and private sectors, resulting eventually in the Joint Marketing Initiative which seeks to bundle all in a single attempt at representing the city in a way that is externally competitive and internally productive.

A target situation exists, therefore, that differs markedly, especially from the period prior to the formation of the Unicity. Between 1996 and 2000 the different municipalities now united in the Unicity often competed with each other, instead of working together. Each would have its own organizations to promote tourism, trade and investment and the rivalry often resulted in the loss of potential investors. Exact figures are not known, but within the city administration many respondents made no effort to hide their frustration

---

[14] Spatially large zones of poverty enveloped by areas of wealth can still be identified (ter Berg 2004; City of Cape Town 2001b).

with this counterproductive situation. Ian Douglas – CEO of Convenco and process manager of the Joint Marketing Initiative – phrases it most eloquently when he states: 'In the past we have had a rather messy, confusing and sometimes conflicting approach to attracting tourists, trade and investors' and 'I know the private sector has been frustrated by this over the past few years' (www.jmi.co.za).

Still, this situation already beheld an improvement from the pre-1996 situation where 49 local authorities could be discerned, each, with its own agenda and supporting organizations – such as chambers of commerce and tourism bureaux. The general trend thus seems to be one of integration of marketing activities and the, for now, final stage in this process is the Joint Marketing Initiative, which is a co-operative effort by the city and provincial administration.

Furthermore, the unison of marketing strategies is greatly helped by efforts such as the finalization of a generalized land-use map in 2001, for which eighteen different categories of land use were identified. And as, in 2003, the city was working on a metropolitan-wide zoning scheme, an even more detailed framework for city planning was on the way.

Strategic direction: the former Cape Town municipality

Cape Town has always been highly centralized. This means that in terms of power and prestige the former Cape Town municipality in general, but the central city in particular, has always retained a dominant position. Thus, it is that the current JMI strategies and the discourses informing these have for a large part been influenced by the strategies and initiatives of this former autonomous municipality.

Here, post-1994, several players could be identified: first of all, the Cape Town administration; secondly, the Cape Town Chamber of Commerce; thirdly, Wesgro; fourthly, the Cape Town Tourism Bureau; and fifthly and last, the Cape Town Central City Partnership. Together, these five formed the core of the city's marketing bureaus. Their strategies were characterized by a strong international focus and an outward perspective, as exemplified by the use of slogans with a definite international ring to them: 'Cape Town, where the world and two oceans meet', 'Cape Town, a World Class City' and 'Cape Town, Gateway to South Africa' to name but a few.

One of the city's prime concerns was to put Cape Town back on the international map, an ambition that came to the fore strongly when the city, rather ambitiously, in 1997 made its bid for the 2004 Olympic Games. And although the games were not won, the bid did generate lots of very welcome publicity. What is of interest, though, is that this was not an isolated attempt, for it fitted perfectly within the overall strategy of promoting Cape Town and the Western Cape as 'a place of great events', a slogan that the Cape Town and Western Cape Tourism bureaux still use on their websites and that represents part of the city's two-legged vision for the tourism sector.

For first of all the city is presented as the perfect international leisure destination, combining a Mediterranean climate, stunning natural beauty and a rich cultural history. Looking to attract the big-spending international tourist, this strategy furthered the realization of up-scale developments, such as the V&A Waterfront and Century City complexes, and also new recreational areas such as the Ratanga Junction theme park partly originate here. And, under the guidance of the central city partnership, the city centre got a facelift. Specifically noteworthy in this case are the pedestrianization of the central city and the renovation of buildings of cultural and historical value. The organizations involved thus tried to cultivate the image of a city of stunning natural and cultural beauty where life is good and people are happy and content. Little, if anything, was said about the apartheid inheritance of a divided city and it is only recently that investors and entrepreneurs have tried to capitalize on this awkward part of the city's past. Township tours and the publicity they generate are a prime example.

The second leg of tourism-related strategy was the previously mentioned promotion of the Western Cape as a place of great events. These events were seen as an important means of presenting the city on a larger stage and as a very effective way to disseminate favourable images of the city to a wide audience. In her study of symbolism in Cape Town, te Velde mentions the importance of the Cape Carnival, a festival which is attracting ever-greater numbers of visitors and is growing in diversity at the same time. Other events staged are an annual cycle tour, all kinds of national and international sports events,[15] wine festivals, etcetera. All in all these festivals contribute to the image of a 'happening' city. Indisputably, the events give expression to the city's diverse character and lend it a sense of bustling activity.

These efforts by the tourism promoting organizations then linked up with the efforts of Wesgro and the Chamber of Commerce as Cape Town was shown as a great place to live and work, an image that was and is of great value when trying to attract investment and investors, since, especially in an era of fierce international competition, an attractive living environment is considered a key asset.

Furthermore, efforts to attract investment tended to focus on niche markets. The film industry, the convention business and the fashion and design industry were specifically targeted. Overall, several organizations

---

[15] The biggest event to date has been the staging of a large part of the 2003 Cricket World Cup.

tried to portray the city as a great place to invest because of its gateway position and the availability of a large and relatively skilled workforce. In comparison with other South African cities, a relatively low crime rate also played an important part in selling the city.

Strategic direction: the JMI initiative

As stated above, a wide range of organizations was responsible for the marketing of the City of Cape Town. Institutions such as the local tourism bureaux, Cape Metropolitan Tourism, Wesgro, the Province and other stakeholders all had a marketing agenda without having an agreed strategy. The purpose of the JMI then, is to co-ordinate the activities of these institutions so as to have a maximum impact. According to the JMI website[16] the main objectives of this initiative are to develop: 'a branding and global marketing strategy for the City of Cape Town and the Cape Region' and 'a co-operative marketing framework supported by an appropriate organization focused on the sectors of Investment Promotion, Trade and Export Promotion, Tourism Promotion, Major Events and Film'. The ultimate goal is to attract more tourists and investors and promote Cape Town as the best place to work, live and play. Thus, in 2001 the JMI got underway with a very clear set of objectives. And although it had, at the time of writing, not yet been able to deliver a finished product in the form of a brand and comprehensive marketing strategy for the City of Cape Town, the outline of these was already becoming evident.

In line with the policies of the former Cape Town municipality and the organizational structure already in place, four separate, though sometimes overlapping, strategic areas can be identified: tourism, events, investment and trade. Furthermore, and in addition to these, the film industry is targeted separately. The Cape Character study (City of Cape Town 2000) provided the basis for the analysis of strengths and weaknesses in these areas and strategies are being developed accordingly. Overall, there is the realization, coming to the fore all throughout JMI-related documents as well as in interviews, that Cape Town should capitalize on its symbolism and iconography as a gateway to the (South) African interior and the entry point towards the rest of (South) Africa.

For tourism, because of its growing importance, its diversity and the high visibility of its impact on city life and structure, the goal is to position Cape Town as a world-class competitor and South Africa's premier conference, incentive, leisure and events destination. Ultimately, the JMI envisages Cape Town to be Africa's premier conference and tourism hub. The emphasis

---

[16] See: www.jmi.co.za

rests on both the location's intrinsic qualities and its favourable gateway and entry point position. According to a draft version of the tourism sector marketing strategy (JMI 2001) the following words should be associated with the experience of visiting Cape Town: freedom, diversity, enjoyment, growth, energy, natural beauty, vibrant, modern, sophisticated, global, trendy, quality, and smart. To achieve this, it is furthermore recognized that promotional efforts should concentrate on some of the city's better known and widely recognized key assets. Thus, the city should, to a large extend, identify itself with its historical position as the place where Nelson Mandela was incarcerated at first, but set free eventually; with the Cape Town city as a cool and trendy place with a cosmopolitan culture and a place of significant beauty at that; with Table Mountain and the Peninsula National Park; and again with its unique geographical location at the point of Africa. Furthermore, the Western Cape should be confidently marketed as a place of cultural diversity, scenic and natural beauty and great wildlife, while at the same time propagating its more mundane qualities such as good shopping and entertainment, good infrastructure and affordability. In a sense, the city and the province are trying to promote the area as a location that has it all – beauty, entertainment, culture – and at a low price at that. The following quote from the website of the Cape Town Tourism Bureau says it all:

> Warmed by the African sun, the city is dominated by a towering, table-shaped mountain, set on a peninsula of soaring, rocky heights and lush valleys ... Immense natural beauty, and the fast pace and bright lights of a great urban centre meld effortlessly here. Firmly positioned on the international map, Cape Town offers a myriad things to do and places to see[17]

(www.cape-town.org)

Then, where events are concerned, the goal as formulated in the draft version of the *Major Events Marketing Strategy* (2001a: 3) is that 'by the year 2010 the City of Cape Town and the Western Cape should be recognized as a unique, world-class events destination delivering real economic and social value to all citizens through the hosting of events'. The objective is to firmly position Cape Town as a potential organizer of large-scale national and international events that are sure to attract international coverage and attention as well as investment. As its key competitive advantage the city sees the favourable reputation it already enjoys in this sector. The experience

---

[17] As with the logos, the following themes can be recognized: nature, climate, cosmopolitan, exciting. In short: fun.

it has with large-scale events such as the African Harvest North Sea Jazz Festival, The Cape Argus Pick and Pay Cycle Tour, The Old Mutual Two Oceans Marathon and the Cape Gourmet Festival should provide the city with the basis to bid successfully for larger national and international events.[18]

Thus, the city's vision, which indeed is very ambitious, links up with an international trend to view large-scale events as important opportunities, not only to increase tourism revenues, but also to promote the city in a wider sense. Events have become an important tool in the marketing of the city. Through the successful bidding for and organizing of ever-larger events, cities are able to promote commercial and economic development and achieve goals such as inner-city regeneration through the use of the events staged as catalysts for development. Cities such as Manchester and Barcelona provide us with successful examples of this strategy. And the fierce competition for the right to organize the Olympic Games shows us the power of this trend.

Moving on to the third sector, that of investment, the slogan seems to be 'Cape Town, a world class business location' (City of Cape Town 2003a), a slogan, which is completely in line with the city's vision as formulated in the draft version of the Investment Promotion Sector Marketing Strategy (JMI 2001c: 2). 'The Western Cape to be the most successful, innovative and competitive economic region in Africa, through its world class, sophisticated manufacturing, agri-industrial, commercial and personal services.' Furthermore, 'to be ranked by global players as the prime investment location in Africa, and a leading location in the Southern hemisphere' is defined as the prime objective of investment-related strategies. It is clear that the city's ambitions reach farther than the national level. Cape Town in all aspects wants to be a world player.

A focused effort has also been undertaken to fare better in modern industries such as media, publishing, printing, and promotion and advertising. Other key sectors have been the traditional strong areas are banking, finance and insurance. And finally, Cape Town has done well to invest in decentralized niche industries such as film and conventions (Wesgro 2001: 19). Thus, it becomes clear that in striving to become a global player Cape Town seeks to move away from more traditional industries, such as manufacturing,[19] and tries to position itself firmly in

---

[18] Just recently South Africa has won the bid for the 2010 Soccer World Cup and Cape Town is bound to be one of its premier sites.
[19] More-traditional industries such as manufacturing and clothing and textiles do remain important, mainly from an employment perspective, but the focus is shifting from the production of very basic goods to the production of more complex and technology intensive

today's frontline industries, such as banking, finance, film and ICT, an approach that is carried by the trust in the city's relatively well-educated workforce[20] and its ability to attract the international elite through the marketing of its favourable living environment.

On a different note, the trading sector, taking into account wholesale, retail, import and export, currently employs over 135,000 people and accounts for about 14 per cent of the Gross Regional Product. But equally significant to its mass is the fact that 40 per cent of this business is in the hands of the informal sector (Top Company's Publishing Pty 2000: 35). The trade sector not only represents an important pillar of the city's and region's economy, but it is also one of the prime areas of grassroots expansion. It is an important sector not only for the city's climb up the international ladder, but also for the achievement of its goal of creating an equitable city.

The JMI vision for this sector: 'for the Western Cape to be a globally connected and efficient trade hub which is supported by an active export culture and which is internationally renowned for high-quality products and services which results in significant trade value growth, trade related job creation, and a growing number of exporting companies' (JMI 'Draft Version of the Trade Promotion sector marketing Strategy' 2001b: 13), is an ambitious one, at the heart of which lies a key concern with quality.

Finally, I turn attention to a sector that undeniably has a growing impact on the cityscape. Speaking from personal experience alone, I ran into film crews almost on a weekly basis and the reports of roads blocked due to (film) shootings appeared in the newspapers almost on a daily basis. In a somewhat comic incident somewhere in March neighbourhood residents actually called in the police after hearing what they thought were the sounds of gang warfare. But rushing to the scene the police encountered just that: a scene. From the newest Vinnie Jones movie that is. And although Mr Jones was arrested a couple of weeks later on different charges, at that time he was not doing any shooting besides shooting a movie. Apparently, though, nobody had deemed it necessary to inform the neighbours.

According to the JMI (draft version of the Film and Television Sector Marketing Strategy 2001d: 4) the goal for this sector is 'to position the region as a top international film location and production destination on par with key international cities, drawing in foreign producers, as well as develop domestic filmmaking capacity'. And even more ambitiously the mission statement says the following: 'To develop the film industry and to

---

products, or, in the case of clothing, from the production of cheap confection towards the production of higher-quality designer clothes.

[20] In comparison with other non-Western cities.

use it as a catalyst to transform the economy of the Western Cape into a globally competitive region' (JMI 2001: 4).

To give them credit, they are well on the way in doing so. Recognized as one of the top five growth industries in the region, the film industry brings in over two billion Rand a year to the region's economy (City of Cape Town 2003b: 5; Wesgro 2001, 2: 44 and 2002, 1: 62). And even more impressively, with eight feature movies shot during the 2000 season, the city is on a par with San Francisco. Evidently, the industry is booming and plays an important part in the proliferation of city images throughout the world.

City types

It thus becomes clear that with the Joint Marketing Initiative the City of Cape Town, through the proliferation of favourable city images, is trying hard to reposition itself on the national and international stage. A conscious marketing strategy and a focus on key sectors lie at the heart of this transitional city's strategy to redefine itself in the post-apartheid era. But the choice of images and focus areas is not solely related to the specific characteristics of the Cape area. Rather, certain themes can be identified that play a central role in post-industrial re-imagining everywhere. City representatives have thus tapped into an international discourse streamlining the marketing efforts of ambitious cities looking to propel themselves forward on the international stage.

In *The Urban Order*, Short (1996) identified four different types of cities with the World City[21] on top – of the world. Within this group there is both competition and collusion. While they need each other to provide 24-hour coverage for international dealings, they often compete for business, investment and status. Their position, though, is relatively stable and secure, which allows them the luxury of a 'metropolitan provincialism' (Short 1996: 429). The belief is that standing in Covent Garden, on Fifth Avenue, or in the Maranouchi district you are right in the middle of things, with the whole world passing through. But the world city group is an exclusive one with New York, London and Tokyo as its only members. Other cities just do not qualify.

Thus, it is below these three that the fiercest competition takes place. And interestingly enough it seems that the strategies used to improve one's position can be divided into a mere three groups (Short 1996: 430), as can be the types of cities accordingly. Apparently, and probably due to the concurring goals and visions, the models underlying any world city strategy are easy to find and limited (Wesgro 2001: 19; Short 1996: 430). Short

---

[21] Extensively studied and written on by Sassen (1991) and Hall (1997) amongst others.

identifies the following groupings of sought for images and accompanying types of cities: wannabee world cities; clean and green; and Look! No more factories. These images are by no means mutually exclusive, though, and the case of Cape Town is exemplary, for all three stereotypes can be found in official communiqués and marketing strategies. But the dominant image portrayed clearly is that of a potential global player, a cosmopolitan city firmly embedded in international networks. The City of Cape Town therefore falls in the category of wannabee world city; an observation of which the validity is unambiguously supported by the language of civic boosterism. Phrases such as: 'Cape Town, a world class city'; 'Cape Town, a world class business location'; or 'Cape Town, in a class of its own' are just some examples of the ambitious terminology used in marketing the city, and speak for themselves. The focus on international markets, as described in the Joint Marketing Initiative, is a further illustration. But the applicability of the concept is most obvious in the pervasiveness of adjectives such as global and international throughout vision and mission statements.

'Our Vision: A smart and globally competitive city …' and
'The city government of Cape Town will build a partnership with all its people to make Cape Town a world-class city …'

(City of Cape Town 2002)

At least in its ambition, Cape Town has clearly outgrown its local and national surroundings. And thus it is not surprising, and provides further proof, that in its evaluations of growth and development strategies, city organizations such as Wesgro, the tourism bureaux and the JMI tend to compare themselves with other international cities such as Rio de Janeiro, Barcelona, Paris, etcetera. Clearly, government officials feel the need to place the city in an international perspective. In interviews, I encountered exactly this mentality. Rather than making comparisons with other South African cities a link would often be made to cities such as Madrid and Montreal. Especially within the planning department, people seem to be oriented towards developments in Northern American and European cities more so than towards developments in South African cities.

Clearly, in the eyes of the administration there is no doubt whatsoever about where this city is going. Sticking with its historical roots the city again aspires to take up its role as a global player. And this ambition is voiced without reservations, as is evident in the quotes above.

Giving further proof, I want to remind the reader of the failed 1997 bid for the 2004 Olympic Games, and the related major events strategy that is currently at the fore of development strategies. Several authors (Sassen and Short amongst others) have already drawn attention to the importance of the bidding for and staging of prestigious international events for those cities that feel they are not quite there yet. In a way, these events are the international fast track towards status and prestige.[22] Hosting a major event such as the Olympic Games thus is one of the key strategies employed by wannabee world cities and as such it is a sign of the Mothercity's aspirations.

I would therefore confidently dub Cape Town a wannabee world city, the more so since the other two categorizations seem to be real long shots. Surely, the manufacturing and clothes and textiles industry have always been strongly represented in Cape Town. But to dub Cape Town a 'factory city' would be to greatly overestimate their importance. And although the city does express its wish to become a 'smart' city it does not truly present itself as a post-industrial city turned around. Quite on the contrary, Cape Town still recognizes the importance of these 'old' industries, although they no longer sit at the centre of development strategies. Thus, although the characteristic processes of gentrification and transformation of old harbour fronts and industrial districts are evident in Cape Town – for example the V&A Waterfront development and the gentrification of parts of the inner city –, they are not used in a 'Look! No more factories' approach. The city just does not define itself in contrast with a 'polluted' industrial past. If, at all, a reverse psychology marketing strategy of a kind can be said to be in use, it would be the positioning of a post-apartheid identity instead of something post-industrial.

As for the third category, Cape Town indeed is very big on environmental issues, but they are not the essence of the message. Surely, the magnificence of the natural surroundings plays an important part in its marketing and the Spatial Development Framework does indeed recognize the necessity of creating a network of green spaces and corridors throughout the city. But these issues are only presented to support the core message that Cape Town truly is a great place to work and live in, with a quality of life measuring up to international standards. The 'green' city is thus seen as a key asset, but not as the sum total. 'Green' indeed is an adjective that fits the city, or at least the image projected, but it is one of many.

---

[22] Barcelona is one of the prime examples of a city where this strategy has paid dividend. Here, the hosting of the Olympic Games was one of the major catalysts behind total city reform and development. Large segments of the city were successfully upgraded and the morale boost and financial injection that resulted from wining the bid and hosting successful games combined to alter exterior perceptions and positively swing inner-city morale.

Superimages

Now, moving from types of cities to the manner in which these global players manifest themselves on the international stage, my concern is, once again, with wannabe world cities, for Cape Town clearly ranks amongst them. As a general rule these cities, in their competition for business, trade and investment, struggle for corporate attention. And because the image is what sells the city, the city in a very real sense becomes the image (Short 1996: 432). Short therefore goes on to identify four main images that cut across the different types and are often superimposed on one another. And as it seems all four bare relevance to the case at hand.

First of these is the 'fun city', where the conspicuous consumption of leisure is central. It is a place where the sky is blue and people are beautiful; a place of smiles and laughs; a place without worries or problems. It is the image portrayed by the tourism industry. The shots of smiling people, sandy beaches and palm trees that fill the pages of brochures and internet sites are exemplary and it is around this image that the city's Tourism Marketing Strategy is largely built. The earlier quote from the Tourism Bureau's website is illustrative.

The building and marketing of leisure centres such as the V&A Waterfront, Century City, Grand West Casino and the Ratanga Junction Theme Park then grounds the theme in the built environment. And the active role that the authorities have played in these diverse projects – informants told me that the Grand West Casino and Century City projects for example were conceived of by local government and privatized only after an intensive tendering process – shows that the city is indeed actively pursuing the reputation of 'fun city'.

Secondly, the 'green city' image also is of great significance for Cape Town's marketing strategy. According to Short, the 'green cities' cultivate the image of a place where 'a nature safely controlled and manicured provides a close though comforting experience with the semi-wild' (Short 1996: 432). Furthermore, he states that water, fresh air and mountain vistas are important icons within this approach. The strong relation between Capetonians and Table Mountain (te Velde 1998) and the iconography of the city, represented through city logos, at the very least strongly suggests that the 'green motive' indeed is a powerful tool for the marketing of the city. But it is more than a mere strategy. Indeed, this might even be the most 'real' part of the city's identity, since the city's natural surroundings in general and Table Mountain in particular provide the city with a unifying identity that is of huge significance to an otherwise divided community.

Thirdly, Cape Town presents itself as a 'Culture City', which Short clarifies as a city that offers 'a manicured and sanitized cultural experience'.

It involves the representation of the city as a cultural festival where everything is on offer, from high-class shopping malls to symphony orchestras. The aim is to project the image of a sophisticated city that is well wired into global culture. The presentation of the Western Cape as the 'Cape of Great Events' immediately springs to mind as does the concern for the city's historic and ethnic culture. The renovation of historic buildings and sites throughout the central city, under the guidance of the Cape Town Heritage Trust, fits in comfortably with this strategy, as does the organization of international cultural events of fame such as the annual North Sea Jazz Festival. Furthermore, the St George's mall and V&A projects have contributed significantly towards the creation of up-scale leisure opportunities. It is here and at the Century City development that a high-class shopping experience is offered.

Finally, Short describes the pluralist theme, where the message is: to 'come and experience the rich mix of different lifestyles'. The colourful brochures and images sent out by the tourism bureaux, Wesgro, the City of Cape Town and others exemplify this approach. Here, Cape Town is shown of as a city with many faces and a rich cultural history and ethnic mix. One only has to surf to a randomly chosen tourism site or to visit the nearest travel agency and flip through the brochures to recognize this image. It is the image of a bustling city for all to enjoy. Life is good and versatile.

Thus, the positive images Short attributed to the developing global city are summed up and shown to be of relevance. But at least one other image has to be mentioned. Since every single person I talked to about my upcoming or past experience of living in Cape Town – except for some of those who had been there themselves – asked me if it was not unsafe. Apparently, and this is something that is widely recognized, cities in South Africa are associated with negative images of high crime levels, the risk of physical harm and a generally high level of insecurity and anxiety, a feeling of unease that is in no way whatsoever lessened by talking to 'white' South Africans, since they often are even more fearful of their own situation than the visiting foreigner. A state of fear that translates into the spontaneous springing up of 'gated communities' and the fencing off of houses. Also, my experience was that many Capetonians will never walk the streets or use the public transport system if they can avoid it in any way. Thus, as Cape Town makes its move on the international stage it will have to work hard not only on solving the very real issues of crime and violence, but also on improving its international reputation. For now, the image of a 'crime city' remains firmly attached, limiting the potential effects of the successful broadcasting of the messages above. It is therefore a very positive and important sign that the City of Cape Town seems to have come to the same realization and that, as a consequence, they are now determined to pursue a promotional campaign that addresses these negative safety and security perceptions. Already in most tourism brochures, on websites and in information leaflets a separate section will be assigned to the issue of safety. The administration in liaison with its citizens is trying hard to lessen the feelings of unease and anxiety.

Discovering the people

Before reaching my conclusions, this concern with people's feelings and experiences leads me to once again touch upon the shift in policy and mentality, already referred to in the discussion of city logos. With the ANC/NNP coalition rise to power in December 2002 came a new way of thinking, but not entirely new but different nonetheless. I say not entirely new because people do not embark on strategies out of the blue. Rather, the turn towards a more inward perspective has been a process that took several years, but only culminated in a definite policy and strategy shift in the summer of 2002/2003.

In the beginning of this article I mentioned that I experienced a general feeling within the corridors of the city's administration that the city had not done enough to inform its citizens and to involve them in the process of guiding their city into the twenty-first century. The Listening campaign that began in February 2003 is one of the first steps in righting this wrong.

But on another level this shift is equally evident. Since 2000 several draft versions of an Integrated Development Approach have been drawn up and after the IDP 2001/2002 was officially accepted by city council, reviews were done on an annual basis. And so one sees the process unfolding, for every document since the first draft has paid more attention to the current socio-economic reality and the vast problems facing the city. Whereas the initial document really was a vision statement, the ensuing publications have become more and more grounded in reality. Eventually, this has lead to a reformulation of the vision statement giving evidence of a more realistic and informed approach. Whereas in the initial drafts and the 2002/2003 version of the Integrative Development Plan (City of Cape Town 2001a, 2002, 2003b) the city envisaged 'to make Cape Town a world-class city, in which the quality of life of every citizen steadily improves', the 2003/2004 version phrases its vision somewhat differently.

VISION FOR THE CITY: Taking into the account the current reality of deepening negative city indicators in areas such as HIV/Aids, TB, infant mortality, crime, homelessness, unemployment and education levels whilst recognising the overwhelming positive attributes of our City, the vision of council is to establish Cape Town as ... a sustainable, dignified, accessible, credible, competent, safe and caring and prosperous city.

(City of Cape Town 2003b: 44)

And this is not simply the extended version of the vision as initially stated. In line with a marked change in attitude the new phraseology represents a deepening concern with the current reality of the city and its inhabitants. It is a sign of a growing concern for the day-to-day lives of Capetonians. And so, instead of starting with an outward and future-oriented perspective, the new vision statement begins with a look at the city's current reality and an evaluation of the quality of live of its inhabitants.

The initial outward orientation is thus completed by a more introvert perspective.

Conclusion

Clearly with the revolutionary events of the early 1990s and the ensuing transformation in 1994 came new opportunities for the City of Cape Town. However, to seize upon these the city suddenly had to compete internationally. And so the city has done. And not without success, either. But nearing the millennium it slowly dawned upon those in power that external marketing and competition alone could not carry the city further. For even in an era where image is everything, representations still need to be rooted in reality for them to convincingly play a role in a city's development. In essence, a minimum level of correspondence between image and the imagined is necessary for the message not to be related to the realm of fiction and fairytales.

Especially in the case of Cape Town, there is a lot of truth in this observation, for at the core of the superimages portrayed rests the behaviour and charisma of the city's population. Fun City, Pluralist City and Culture City are all images that relate to an interesting and diverse experience; an experience that is for the greater part formed in interaction with the local population. All these experiences are staged in the city and those dwelling in it must thus, necessarily, be involved. For the image to be carried over into experience, the co-operation of Capetonians themselves seems a sine qua non. It is the people that have to exude this message of diversity, fun and culture, since it is people that enliven a place. In order for adjectives such as freedom, diversity, enjoyment, growth, energy, natural beauty, vibrant, modern, sophisticated, global, and trendy to stick to the city it is the city's inhabitants that have to convincingly portray this image.

Since not all people are good actors, it is therefore of prime concern to restore people's faith in this message. And thus it is uncanny to see that Cape Town has not in an earlier stage realized to the full extent the importance of informing and involving the people on more than a project basis. For, if the momentum generated through the successful external marketing of the city is to be capitalized upon, the expectations of outsiders cannot be disappointed. Internal and external representations and perceptions of the city are inseparably entwined.

It is therefore interesting to see that the development in logo's and policy documents points to a new stance of the administration. As illustrated in the previous sections, there first of all is the realization that the communication – two-way – with the populace has to be intensified. And secondly, the people and the diversity of the city's environment are now truly recognized as important assets. On the other hand, as the quote in the previous section shows, the city is not ignoring negative city indicators such as crime and HIV. It is rather realizing that, notwithstanding all these problems, part of the apartheid legacy might also be seen as an asset, provided that the poor socio-economic situation is improved upon and the racist divide of the city all but vanishes.

In a sense, the city has embarked upon a strategy to restore pride to all its communities and in so doing it hope to boost morale, something that in turn will reflect positively upon city development. In order for people to be willing to embrace further development they have to believe that a positive spin-off will befall them. The strategy – image is everything! – used internationally, wherein the package was the message, is thus transported to the local level. Through a re-imagining of the city, the administration is trying to change attitudes and perspectives and ultimately the goal is to change reality, for the imagined is supposed to match the image. Following the logic of international competition, the fine line between portrait and portrayed is supposed to disappear.

Theoretically, a couple of observations can be made from this case study. First of all, it is of interest that the literature on transitions from the industrial to the post-industrial city seems to bear relevance to a broader field of research. The case of Cape Town illustrates how the types of cities and superimages identified in research on the post-industrial city apply to any major city with a strong external focus going through a transformation.

Furthermore, the pre-occupation of modern cities with the way they present themselves within a highly competitive international playing field seems to bring together insights from the field of urban symbolism, world city literature and the discourse on 'the city as text'. The study of the entanglement of imagined environments and the material determinants of public life should therefore provide fruitful insights into the modern city.

The development of the city logo and city marketing strategies in relation to the material and socio-cultural environment has thus become an interesting field of study.

And finally it has to be recognized that for image-building strategies to work internal marketing efforts may be just as important as external marketing efforts.

References

Berg, R. ter (2004) *The City as Text: Images, Space and Symbols in Cape Town, South Africa*. Leiden. [MA thesis, Leiden University.]

City of Cape Town (November 2000) *Cape Character Study*. Cape Town: Planning and Development Directorate, City of Cape Town.

—— (2001a) *Interim Integrated Development Plan 2001/2002: At the Beginning of the Transformation*. [www.capetown.gov.za/idp/pdf/part1.IIDP.pdf]

—— (2001b) *State of Environment Report Year 4*. Cape Town: City of Cape Town.

—— (2002) *Integrated Development Plan 2002/2003*. Cape Town: City of Cape Town.

—— (2003a) *Partnership in Delivery*. Cape Town. [Publication is part of the Listening Campaign.]

—— (2003b) *Integrated Development Plan 2003-2004*. Cape Town: City of Cape Town.

Donald, J. (1992) 'Metropolis: The city as text.' In: R. Bocock and K. Thompson (eds.), *Social and Cultural Forms of Modernity*, pp. 417–70. Cambridge: Polity Press.

Hall, P. (1997) *Megacities, World Cities and Global Cities*. Lezing in het kader van de eerste megacities lecture. [http://www.megacities.nl/lecture hall]

Joint Marketing Initiative (2001) *Tourism Marketing Strategy, draft version*. Cape Town. [Internal document by JMI sector team headed by Dr Mike Fabricius.]

—— (2001a) *Major Events Marketing Strategy, draft version*. Cape Town. [Internal document by JMI sector team headed by Pat Lennox.]

—— (2001b) *Trade Promotions Sector Marketing Strategy, draft version*. Cape Town. [Internal document by JMI sector team headed by Rae Wolpe.]

—— (2001c) *Investment Promotion Sector Marketing Strategy, draft version*. Cape Town. [Internal document by JMI sector team headed by R. Hein.]

—— (2001d) *Film and Television Sector Marketing Strategy, draft version*. Cape Town. [Internal document by JMI sector team headed by P. Mseleku.]

Nas, P.J.M. (1984) 'Settlements as Symbols: The Indonesian Town as a Field of Anthropological Study.' In: P.E. Josselin de Jong (ed.), *Unity in Diversity*, pp. 127–42. Dordrecht: Foris Publications.

Nas, P.J.M. (ed.) (1993), *Urban Symbolism*. Leiden: E.J. Brill.

Sassen, S. (1991) *The Global City: New York, London, Tokyo*. Princeton: Princeton University Press.

Short, J.R. (1996) *The Urban Order*. Cambridge: Blackwell.

Top Company's Publishing Pty (2000) *South Africa's Top 300 Companies: Cape Regions*. Cape Town: Top Company's Publishing Pty.

Velde, M.N. te (1998) *Between Devil's Peak and Lion's Head: Urban Symbolism in Cape Town, South Africa*. Leiden. [MA Thesis, Leiden University.]

Wesgro (2001) *Cape Africa*. Cape Town: Wesgro. [Volume 2.]

—— (2002) *Cape Africa*. Cape Town: Wesgro. [Volume 1.]

Chapter 10

BRUSSELS AS A SYMBOL
Controversial images of the capital of Europe

Alexandra Bitusikova

The symbolic significance of a city used to be described by early urban sociologists and anthropologists in contrast to nature and countryside (Redfield 1947; Wirth 1938). The city has been considered the greatest symbol of civilization, the biggest achievement of human society, centre of learning and communication, meeting point, space for activity and mobility. On the contrary, the countryside has been usually defined as a place of backwardness, ignorance and limitation (e.g. see Williams's historic analysis of literary representations of the countryside and the city, 1973). The negative descriptions of the city present it as a place of noise, ambition, decadence, anonymity, loneliness, impersonal relations, crime, drugs and prostitutes, which is in contrast to the rural environment as a place of peace, harmony, innocence, isolation, solidarity, homogeneity and strong family ties (Redfield 1947; Williams 1973; Wirth 1964). In addition to this general symbolism of a city, particular cities develop and produce their own special symbols that contribute to constructing urban identity of the inhabitants on the one hand, and to creating outside image of the city on the other.

Urban identity is an important phenomenon in the process of production of symbols, and vice versa – symbols are important in the process of constructing urban identity. Experience of being and living in the city, eating in restaurants, walking in streets and squares, bargaining in flea markets or enjoying art galleries provides many opportunities for identity formation (Zukin 1998). There are several factors influencing urban identity: symbols (official and unofficial, formal and informal, etc.); natural and geographical phenomena; language; urban events and rituals; memories, emotions, images and stereotypes. It is mainly places of integration that often become a symbol of identification of the inhabitants with their city. According to Edwin Eames and Judith G. Goode (1977), units of integration are phenomena bringing various groups of heterogeneous urban population together and supporting communication and interaction of city inhabitants. These are either space localized (public spaces such as squares or streets, restaurants, pubs, markets, public institutions, etc.) or time oriented and localized (urban rituals such as festivals, celebrations, political and sports events). All these public spaces and events have an impact on production of symbols – both through the revitalization of old city symbols, and the creation of new ones. In many European cities we can witness a revival of symbolic events and objects with the aim to strengthen urban identity, but also to attract more visitors or investors to the city (as in many Central and East European cities in the last decade; e.g. Bitusikova 1998). It is municipalities which are main initiators of the revitalization of urban identities by reconstructing city centres, organizing urban rituals, creating or reinventing urban traditions. Michelle de la Pradelle argues and asks whether this newly constructed, directed and organized identity has a real impact or importance for the city inhabitants (De la Pradelle 1996: 192). According to Kevin Lynch's theory, production and understanding of structure (space, object or event) opens an opportunity for urban inhabitants to give urban place a meaning and to construct their urban identity (1960).

If we follow the definition of a symbol as a relationship between the object (statue, building, flag, song, etc.) and a meaning outside the object (Colombijn, this volume), we usually mean public objects and their public meanings that are more or less accepted by a number of people or even by the majority (e.g. Eiffel Tower as symbol of Paris or Big Ben as a symbol of London). This has also been a part of Lynch's observation of public forms of meaning. In addition to this kind of symbolism of 'public' domain, we can also talk about symbolism which is the overlap of many individual images (Lynch 1960: 46), based on personal and private experience. Lynch argues that people hold in their minds images of the city essential to their experience and interaction with it. Every individual living in a city or passing through a city gives a special meaning to a particular place or object (places or objects) that brings strong associations or memories. Polish architect Pawlowska (Pawlowska 1998) metaphorically describes these places as 'magic places' – *genia loci* that attract people by their atmosphere, provoking positive memories and emotions. These can be certain streets, squares, parks, monuments, cafes or restaurants, e.g. public places with 'a soul', smell, colour, taste, light or sound (silence); favourite places of social interaction, communication or relaxation. These places are rarely mentioned in official tour guides as symbols or landmarks, but for the inhabitants they are closely connected with a socio-cultural image of their city and have a strong influence on urban identity, often stronger than 'official' and 'formal' symbols.

Lefebvre develops the idea of symbolic aspects of people's images even further. He describes 'representations of space' (used by planners who lack sensitivity towards everyday spatial practices of the people) and 'representational spaces' that incorporate symbolic meanings (Lefebvre 1991).

Authors studying urban symbolism talk about different categories of symbols: profane and sacred, formal and informal, private and public, natural or cultural, soft and hard, open and closed, etc. (Nas, Jaffe and Samuels, this volume). In this contribution, I will try to study urban symbolism in the city of Brussels from two perspectives reflecting the city's struggle for its identity: the symbols of Brussels as a national capital and the symbols of Brussels as the capital of Europe.

Welcome to Brussels – the heart of Europe

These words are the first to be seen when one gets off the plane at the Brussels airport. It seems that Brussels is one of many hearts of Europe, but would like to be the only and the real one. There is a long way to go to achieve this aim. The capital of Belgium, as well as the uncrowned capital of Europe, is an interesting laboratory for study of urban symbolism. The city's identity (as self-perception) and image (as perception and reflection by others) are most controversial due to its history and place both within Belgium and the European Union. Brussels seems to struggle between the two positions as a national capital and a political centre of Europe, which is reflected also in production of symbols or lack of it. In addition to that, due to its prevailing francophone character (e.g. language, restaurants and food culture) the city lives in the shadow of Paris, often reflected in people's cynical comment: 'The best thing about Brussels is that it is only an hour and half far from Paris.' Brussels is also known as the most boring city in Europe, 'more boring than Brussels sprouts'.

The negative image of the city is strongly supported by media and writers. Before I moved to Brussels, I read the book *Neither Here nor There: Travels in Europe* written by Bill Bryson, a famous author of travel books. His presentation of Brussels had negatively influenced my image of the city before I even visited it:

Brussels is a seriously ugly place, full of wet litter, boulevards like freeways and muddy building sites. It is a city of grey offices and faceless office workers, the briefcase capital of Europe. It has fewer parks than any city I can think of, and almost no other features to commend it – no castle on a hill, no mountainous cathedral, no street of singularly elegant shops, no backdrop of snowy peaks, no fairy-lighted seafront. It doesn't even have a river. How can a city not at least have a river?

(Bryson 1991: 65)

Such a description of the city written by a bestselling author of travel books has a significant impact on the outside image of the city and construction of its heterostereotypes (perception of the city by the others). However, it is important to stress that despite all negative or neutral images of the city (whether fair or not), there is a lot in Brussels to be discovered. Brussels inhabitants say that the most beautiful features and symbols of Brussels are hidden and one has to find them for himself/herself. It is obvious that Bill Bryson failed at this task.

Symbols of Brussels as a national capital
Nature
For the people living in the ancient times, nature was full of symbols and hidden meanings. Natural places such as forests, hills, rivers, valleys, caves, lakes, etc. represented something which was beyond the world, beyond understanding. This sacred and respectful attitude towards nature and natural objects survived for a long time in rural communities which were highly dependent on nature and its resources.

In urban environment nature plays a different role. Each city has some natural islands surrounded by man-made places or natural panorama, which may become symbols of the city, such as Table Mountain in Cape Town, the Seine and Bois de Boulogne in Paris, Hyde Park and the Thames in London or Palatine Hill and the Tiber River in Rome. Unlike most European cities that have at least one characteristic natural feature, e.g. a hill, a river or a forest, Brussels lacks any of these. If you ask any inhabitant or visitor about images and symbols of Brussels, nature appears on the bottom of the list. Brussels does not have any significant natural landmarks, no mountains and no river. In fact, there was once a River Senne and a smaller River Maelbeek in Brussels, but they were both covered up and became a victim of the town planning in the nineteenth century.

However, despite the lack of natural symbolism, Brussels is one of the greenest capitals in Europe. (In this case Bill Bryson was not right in his criticism.) About 15 per cent of city space is covered by parks and forests (e.g. Forêt de Soignes and Bois de la Cambre as the biggest ones). The Royal Park, originally a forest with little hills and valleys, was turned into a new park in classical style under the rule of the Habsburgs. It became a social place where the music was played and people could walk, sit, eat and relax. In 1830, the park became the battlefield and the cradle of the Belgian independence. In spite of this symbolic significance in Belgian history, the Royal Park these days is no longer a lively public space. It serves as a place of relaxation, but does not provide space for many ceremonies and carries no direct symbolic messages for the present inhabitants.

In addition to large green parks and forests, there are dozens of tiny, nicely laid-out parks all over the city, often hidden in residential neighbourhoods. They usually provide homes for statues and monuments. During the weekends, city parks and woods come alive, attracting various groups of multicultural population. All green areas are neutral in relation to any groups of population. They do not have a significant place in the perception of the city by its inhabitants, but they may play a role in private symbolic meaning of public places, and they contribute to integrating of urban population and strengthening urban identity.

Nature does not play an important role in the image and symbolism of the city. It is obvious that Brussels has a potential to turn some of its green areas into symbolic features by giving them a special meaning, but neither communes (city district local governments with high authorities) nor inhabitants have initiated any significant activities of this kind.

History reflected in Brussels symbols
The history of Brussels started in the tenth century when the fortress on the island of Saint Gery was built. The economic power of the city increased in the Middle Ages from the growing importance of trade guilds controlled by dukes and duchesses. The impressive Grand Place (Figure 1), often described as the most beautiful square in the world, leaves no doubt about the wealth of the guilds. Today, it is packed with tourists, and local people (although they appreciate the beauty of this space) prefer to spend their time in other areas. The Grand Place has become one of the most powerful symbols of the city as the national capital. With a chain of medieval side streets full of restaurants with seafood (Ilot Sacré, Holy Isle) it constitutes the heart of the city where most of urban events take place (rituals, concerts, parades, festivals, flower show, etc.).

Figure 1 Grand Place: The main square and pride of Brussels
(Photograph: Jonathan Wootliff, 2002).

Situated in the centre of Western Europe, Brussels came under different rulers – the Austrian Habsburgs, the French Charles V and Louis XIV, and the Dutch William I. As a result, the architecture of the old city is a mix of Teutonic and Latin cultures, a cocktail of different styles merging in a special harmony that can be described as Belgian.

Brussels is one of the youngest capitals in Europe. It became the capital of Belgium after the country gained independence in 1830. It was a victory of the coalition of liberals and Catholics, Flemish and Walloons and two languages – Flemish and French. Although Brussels was and later became also officially bilingual, it is *the language* that was always a dividing factor within the Brussels population. In the eighteenth century Brussels was mainly Dutch-speaking with a small stratum of French-speaking nobility. The expansion of the Francophone groups in the first half of the nineteenth century had an impact on linguistic behaviour of the middle classes in Brussels (Van Velthoven 1987: 21–7). French was used by the upper classes and it was a language of administration, courts and higher education. Flemish was considered a language of the lower classes. To achieve prestige and higher status, Brussels Flemish adopted the upper-middle-class language policy and looked down on their Flemish origin and language, which had an impact on the status of Flemish and the whole language and cultural situation in Brussels until the twentieth century (Van Velthoven 1987: 27–8). Language tensions have survived to the present, especially in suburbs of the city where Flemish and Walloonian communes and regions meet.

The official bilingual character of Brussels is best reflected in street names written in two languages, which may be very confusing for newcomers, often finding two completely different names of a street. In addition to that, Brussels has its own language, Marollien, spoken now only by a small number of old 'Bruxellois' in the Marolles district. Marollien is a Dutch-based dialect, which contains also words from French, Spanish, German, English and Yiddish. In 2004, a campaign was launched to make Marollien one of the official EU languages by 2006. Old and forgotten streets in the Marolles district still carry names in three languages: French, Flemish and Marollien.

Despite the official bilingualism of the city and ethno-linguistic confrontations and struggles between French and Flemish, multilingualism has become a strong feature of the present city atmosphere. Since the nineteenth century, Brussels attracted significant numbers of immigrants from neighbouring countries, in the second quarter of the twentieth century from Spain and Italy, and in the late 1960s it saw a massive immigration from Morocco and Turkey. EU civil servants also contribute to a multilingual and multicultural character of Brussels (Witte and Beardsmore 1987: 11–12). However, despite the official bilingual status and existing multilingual and multicultural character, French remains dominant in everyday street communication as well as on posters, billboards and unofficial documents.

The turbulent history of the *'carrefour de l'Europe'* left traces on the architectural face of Brussels. The nineteenth and twentieth centuries witnessed the most dramatic changes of the city – both for the better and the worse. In the second half of the nineteenth century, most impressive avenues and buildings were built under the rule of Leopold II (1865–1909), the Belgian King with a big ambition and appetite for a powerful colonial Belgian empire. The architecture of this period (Avenue Louise, Arc de Triomphe and Parc de Cinquantenaire; Palais de Justice; African Museum in Tervuren; Jardins Botanique, Royal Palace, etc.) is represented by a heavy monumental, but coherent style of impressive buildings in several districts and streets of Brussels. In spite its monumentalism, it is not very much appreciated today by those who know Belgian history. These massive buildings symbolize Leopold II's arrogance and obsession with power, and they reflect his colonial policy of extracting maximum profit from the colony of Congo regardless of vast human and environmental costs. Imperial Brussels of the second half of the nineteenth century reminds everyone of the cruelties and atrocities committed in Congo under Leopold II. The architecture of this period is an expression of state power and ideology, the same way as Blockmans describes Berlin, Moscow and Istanbul (Blockmans 2003).

The end of the nineteenth century witnessed an amazing bloom of a new style of architecture that has marked Brussels strongly: Art Nouveau. Several talented architects, with Victor Horta as the most famous one, designed a large number of buildings inspired by plants and flowers. Art Nouveau style suited the individualistic character of the Belgians and developed without any co-ordination in town-planning. As a consequence, there are many streets in Brussels where each house shows a completely different style, shape, height, windows, gates, ornaments and colour, but the result of this diversity is a charming, extravagant and original mixture, very much characteristic of Brussels (Figure 2). Art Nouveau became popular mainly among socialists and free-thinkers. The Belgian Workers' Party gave its approval of the style as a concrete illustration of Marxist ideology, while the Catholics condemned it as godless extravagance (Brussels Guide 1998: 29–30).

Figure 2 Typical Brussels street with diverse houses built in Art Nouveau style
(Photograph: Jonathan Wootliff, 2004).

The *town planning* in the nineteenth- and twentieth-century Brussels can be described as a series of urban disasters and catastrophes. A large number of historic buildings that could not be used for the purpose of modern life and work or turned into offices were destroyed or neglected. Brussels authorities could not show less regard for its heritage. Most of Victor Horta's famous masterpieces became a victim of such an attitude. Not surprisingly, 'bruxellization' is a term used by urbanists and architects worldwide as a synonym as well as a symbol of urban destruction.

Brussels may have a wide range of architectural styles, beautiful squares and buildings, but its most famous symbol has nothing to do with any spectacular architecture or grandioso monuments. The quintessential Brussels symbol provokes at first either cynical comments or surprise and disappointment. Manneken Pis, the tiny seventeenth-century bronze statue of a pissing boy, with an unclear history, is locally a beloved icon, and a ridiculed symbol of the city (Figure 3).

It attracts thousands of tourists who, when seeing it, cannot believe how small it is. It is hard to see any special meaning or message behind this statue, but there are several urban legends about Manneken Pis. One says that he was a boy who extinguished a fire and saved the city, which could be one of several possible explanations of such an adoration and protection. The sculpture has been stolen and found several times, and always hidden to protect it against bombs during the wars. Manneken Pis is a central part of urban rituals.

He 'owns' more than 600 costumes (he received the first one in 1698 from the governor of the Austrian Netherlands) and it is supposed to be a honour for a government, a famous individual or various interest groups to donate a costume for him. The act of dressing Manneken Pis up has become a ritual itself. There is only one man, the head of ceremony, who is allowed to dress the sculpture and it happens in the presence of the costume donator. At these special occasions, Manneken Pis pees beer for the great pleasure or disgust of the crowds watching the ceremony.

To be politically correct in the age of gender equality, in 1985 Manneken Pis was given a counterpart – Janneken Pis, a pissing girl figure that is considered vulgar and ugly by almost every visitor. The final proof of the mischievous spirit and weird sense of humour of the Bruxellois is a statue of an anonymous pissing dog in the Ste Catherine area (Figure 4). It is not as famous as Manneken and Janneken, but it deserves to be mentioned and seen.

Figure 3 Manneken Pis: The famous symbol of Brussels
(Photograph: Jonathan Wootliff, 2005).

Brussels is home to a large number of statues, which represent various historical events, national figures or themes. It seems that city inhabitants like their statues. It is reflected in their creative production of symbolic meanings, which is often very informal and witty. Statues sometimes receive various accessories – either a piece of clothing, glasses or hat, which makes them more alive – they become real objects of human communication and contact.

A famous symbol of modern Brussels is a futuristic model of the nine molecules of an atom, the Atomium (Figure 5). It was constructed in steel and aluminium by André Waterkyen for the World Fair in 1958 and was supposed to be destroyed afterwards but, due to its success, surprisingly survived and became a modern symbol of the city, mentioned in every tour guide and photographed by every tourist. This story demonstrates that production of symbols does not necessarily need to be planned, but may be unpredictable (the same way as the Eiffel Tower was built for the International Exhibition in 1889 and was welcomed with violent hostility and criticism by most Parisians).

Figure 4 Pissing dog: Another proof of mischievous humour of the 'Bruxellois' (Photograph: Jonathan Wootliff, 2004).

Atomium, together with the Basilique Nationale du Sacré Coeur (one of the biggest churches in the world) and the Hotel de Ville (with a 100-metre tower at the Grand Place) are the dominant architectural structures that make the characteristic skyline of Brussels.

Symbols of Brussels as the capital of Europe

Brussels is a city chronically criticized and misrepresented by many journalists, observers, politicians or any individual citizens of the European Union. In terms of urban symbolism, the city itself plays the role of a symbol all over Europe – mainly in political and media jargon. 'Brussels has decided about new rules', 'Brussels is responsible', 'Brussels dictates' – these subtitles are very familiar to every citizen in any EU Member State. Usually they carry a negative connotation of power and decision-making outside individual states of the EU. Brussels remains a main symbol of bureaucracy and corruption. Any decision made in Brussels is interpreted by national politicians in the EU Member States according to its nature – if it is an unpopular decision, 'Brussels' is blamed; if it is a good decision, national

Figure 5 Atomium: The modern symbol of Brussels and a dominant feature of the city skyline
(Photograph: Jonathan Wootliff, 2005).

politicians take the glory. For this reason, Brussels carries a negative heterostereotype, and is very often negatively perceived not only by outside observers (including many Belgian provincial inhabitants), but also by the native 'Bruxellois' who perceive the presence of foreigners working for the European institutions as intrusive.

The vague and negative image of Brussels as the capital of Europe is a well-known phenomenon. European citizens did not choose Brussels as their capital and therefore it does not play a significant role in their European identity formation. At the Nice Summit in December 2000, European leaders decided that the role of Brussels as the capital of Europe had to be strengthened (Brussels 2001) by identifying and creating symbols, messages and appropriate forms of communication that would be visually and physically expressed in everyday life and practice. A high-level group of intellectuals was invited in 2000 by Romano Prodi (President of the European Commission in 1999–2004) and Guy Verhofstadt (Belgian Prime Minister) to discuss the issue. The key points of the discussions led to many 'shoulds', but not many 'will dos':

1. Brussels should become a symbolic centre of diversities where all issues about diversity, tolerance, integration, globalization, religion and multilingualism are discussed and taught at various events and perhaps at a multilingual open university;
2. Brussels should become a capital of European culture by organizing regular presentations of different European regions, comparative literature book fairs, and multinational cultural festivals, art exhibitions, multilingual theatre, etc.;
3. Brussels should play a different role than national capitals and serve as a networking centre;
4. Improving Brussels' role of the capital of Europe has to be closely connected through the physical substance and buildings of the European institutions (Brussels 2001).

The European neighbourhood (*Le Quartier Européen*) with dozens (or hundreds?) of administrative buildings is one of the main nightmares of the image of Brussels. The central part of the city is filled with large, ugly buildings that are workplaces for thousands of Eurocrats. Professionals, Eurocrats (eurocrats *fonctionnaires* and young *stagiares*, called also Eurostars), coming from all member states, candidate and associated countries of the EU, as well as the people from many multinational companies, known as expats, represent about 15 per cent of city inhabitants (Favell 2001) and they are the main power and energy contributing to a multinational and multicultural character of the city. In addition to them, there is a large population of the North African (mainly Moroccan) and Turkish immigrants who complete the mosaic of a heterogeneous multi-ethnic and multicultural image of the city. Altogether, foreigners represent about 30 per cent of the whole Brussels population (Brussels 2001). Professional Eurocrats and expats stay in Brussels for an average of four-and-a-half years before moving on (*Bulletin* 25 March 2004). During the weekends they often commute to their places of origin and permanent residency, which means that they do not have time, energy and willingness to build a relationship and a feeling of identification with the city.

European symbolism in Brussels is mainly reflected through the European Quarter (*Le Quartier Européen*). A number of streets and squares are named after the founders or leading figures of the European Union with Schuman Square (*Rond Point Schuman*) as the centre of the European quarter (Figure 6). Most of buildings in this part are designated with the EU flag (Figure 7). Their impersonal, cold image and often enormous size are a clear expression of political power and bureaucracy, and reflect the distance between political power and 'normal citizens'. The quarter is only alive during the working days and hours; in the evenings and during the weekends there is no reason to visit this part – everything is closed. The European quarter stands out like a foreign, separate part of the city. Its buildings appear just as an office space, isolated from other forms of symbolic representation and communication (Brussels 2001). The idea of re-inhabiting and revitalising life in the quarter seems to be one of the ways of bringing European Brussels closer to its people. So far it has remained only a manifestation of words, but not actions. Even the European Parliament that should represent a bit of each Member State in a soft and welcoming way is only a magnificent but isolated, cold building with almost no symbols of European states (except for the flags), cultures, languages; no symbols of diversity and unity is visually and physically present in this building. There are plans to build a European park around the parliament, but it is a long process to find a consensus among all partners (political representatives, architects and town planners, neighbouring communes and inhabitants).

Figure 6 Schuman Square, the bureaucratic heart of the EU capital
(Photograph: Jonathan Wootliff, 2004).

234

The production of European symbols in Brussels has been one of the key issues during the debates of the high-level group of experts on the role of Brussels as the capital of Europe (among other seminars and conferences). Romano Prodi criticized that there was no significant symbol of Europe in Brussels that could be comparable with other, older symbols (Brussels 2001: 23). The focus should be on giving buildings and monuments a meaning. European Brussels has to be presented both by soft, light symbols (cultural events, networking and communication) and by hard symbols (buildings, monuments, statues, etc.).

Social production of European symbols in Brussels is closely connected with the problem of European identity. The question is whether European identity exists, and if it does, what does it mean? I think European identity is a construction of multiple identities and it cannot be persuaded only by creating new symbols. Each individual has many identities that are contextual – in various situations one feels and identifies himself/herself differently. European identity mainly appears or works in a non-European context: we feel Europeans in the US or Asia when confronted with different non-European cultures and ways of life. Building European identity seems to be even more difficult than creating European symbols, and it should not be artificially persuaded, but a part of successive identification of each individual with integrated Europe while keeping other identities.

Despite the lack or negative presence of European symbols in Brussels, the city is a vibrant multicultural Eurocity with high level of productive economic and political activities and sociability, and good quality of life. However, if it wants to gain the hearts of citizens of Europe, European and local authorities, architects and town planners have to find more effective ways of communication and consensus. The latest results of the European Parliament elections with the lowest voter turnout ever (June 2004) demonstrate that European institutions and their home town of Brussels have to improve their image across Europe by making them more attractive, more welcoming and open to all citizens.

Conclusion

In this article I tried to outline – in a descriptive way – the main problems of urban symbolism in Brussels, the national capital of Belgium and the capital of Europe. Brussels struggles between these two positions, which is further reflected in a lack of the city's integrity. As expressed by a respondent:

Figure 7 European Quarter in Brussels (Photo: Jonathan Wootliff, 2004).

'Brussels is not an integrated city. It does not have a soul. No one really owns it.' This statement is not surprising if we take into account the number of the city and European authorities:

1. Brussels is the seat of the Belgian federal government and parliament; the Walloonian government and parliament; the Flemish government and parliament; the Brussels government and parliament; representations of three language communities (French, Flemish and German) and nineteen communes (city districts) with their own identities and strong administrative autonomies, which allow them to have separate development policies in many areas – town planning, education, healthcare, police, etc. No wonder that Brussels inhabitants find the city disintegrated. Vast national, regional and communal bureaucracy and language diversity in the capital of Belgium results in lack of coordination among various levels of administration and offices. Consequently, this situation is reflected in the production of symbols and their public meaning, which is disintegrated, and does not have a clear conception.

2. In addition to the complex national administration, the city is home for the European Union – the European Commission, Council of Ministers and European Parliament (shared with Strasbourg), the headquarters of NATO and the seat of hundreds of other institutions related to European and NATO policies. This gives Brussels a negative label of the capital of bureaucracy. Brussels suffers from being used as a synonym for all the down sides of European administration. The European Union has so far failed to improve the European image of the city. Symbols – both soft ones (such as organizing European events or ritualized celebrations contributing to networking and communication outside European political agendas and cold administrative buildings), and hard ones (such as buildings, parks, monuments or statues that would reflect European cultural richness and diversity) could help to build the image of Brussels as the soft, welcoming capital of Europe instead of the image of the seat of power, bureaucracy and corruption.

Production of symbols and meaning is closely connected with constructing and strengthening identity. If the production does not have clear objectives and directions, it is more difficult for the inhabitants to build their relationship with the city. Urban identity of the Brussels inhabitants does not have strong foundations on which to build. The disintegration of the city due to its multiple administration, and a large number of mobile professionals who stay in the city only for a while, do not provide good conditions for building strong and stable identity.

Study of urban symbolism in Brussels opens interesting challenges for a researcher. I only touched on them briefly, but I hope that future in-depth research will contribute not only to basic data, but also to the theory of urban symbolic ecology.

References

Bitusikova, A. (1998) 'Faktory formovania urbannej identiy v sucasnosti (Factors of forming urban identity).' In: *Acta Universitatis Matthiae Belii 2*, pp. 274–84. Banska Bystrica: Univerzita Mateja Bela, Fakulta Humanitnych Vied.

Blockmans, W.P. (2003) 'Reshaping cities: The staging of political transformation.' *Journal of Urban History*, vol. 30, no. 1, pp. 7–20.

Brussels (2001) *Brussels, Capital of Europe: Final Report of the Group of Policy Advisers (GOPA)*. Brussels: European Commission.

Brussels Guide (1998) *Time Out*. London: Penguin.

Bulletin.

Bryson, B. (1991) *Neither Here nor There: Travels in Europe*. London: Black Swan Books.

De la Pradelle, M. (1996) 'Nekolik poznamek k urbanni antropologii (A few notes on urban anthropology).' *Cesky Lid*, vol. 83, no. 3, pp. 189–95.

Eames, E. and J.G. Goode (1977) *Anthropology of the City: An Introduction to Urban Anthropology.* New Jersey: Prentice-Hall.

Favell, A. (2001) *Free Movers in Brussels: A Report on the Participation and Integration of European Professionals in the City.* Brussels: IPSOM Working Paper.

Lefebvre, H. (1991) *The Production of Space.* Oxford: Blackwell.

Lynch, K. (1960) *The Image of the City.* Cambridge: MIT Press.

Pawlowska, K. (1998) 'The perception of cultural urban landscape.' In: P. Jancura (ed.), *Krajina, Clovek, Kultura* (Country, man, culture), pp. 31–4. Banska Bystrica: SAZP.

Redfield, R. (1947) The folk society. *American Journal of Sociology*, vol. 52, pp. 53–73.

Van Velthoven, H. (1987) 'Historical aspects: The process of language shift in Brussels: Historical background and mechanisms.' In: E. Witte and H.B. Beardsmore (eds), *The Interdisciplinary Study of Urban Bilingualism in Brussels*, pp. 15–45. Avon: Multilingual Matters.

Williams, R. (1973) *The Country and the City.* New York: Oxford University Press.

Wirth, L. (1938) 'Urbanism as a way of life.' *American Journal of Sociology*, vol. 44, July, pp. 1–24.

—— (1964) *On Cities and Social Life.* Chicago: The University Press.

Witte, E. and H.B. Beardsmore (eds.) (1987) *The Interdisciplinary Study of Urban Bilingualism in Brussels.* Avon: Multilingual Matters.

Zukin, S. (1998) *The Cultures of Cities.* Oxford, Massachusetts: Blackwell.

Chapter 11

THE USE OF SYMBOLS IN CITY MARKETING
The case of Delft

Th.B.J. Noordman

Introduction

The theory of urban symbolism is trying to cross its present boundaries. One possible way to realize this is to look for practical applications outside of cultural anthropology. City marketing makes intensive use of symbols as landmarks and icons, and without much ado. Could it be interested in a theory of applied urban symbolism?

In this article I will try to answer this question and in so doing outline such an applied theory of urban symbolism. A research project on the identity of the city of Delft in the Netherlands (Risbo 2004) forms the basis for my reflections. We recommended several changes in the city's marketing policy to the mayor and aldermen of Delft. Symbols occupied a prominent position in these recommendations.

Delft is a very interesting city in terms of symbols and rituals. It belongs to the *genus* of Dutch cities linked to the Dutch royal family: Leiden, Amsterdam and Delft. Leiden is where members of the royal family go to study, Amsterdam is the capital city where members of the royal family get married and Delft is the city where they will be buried. Moreover, in the seventeenth century – the famous Golden Age − Delft was the third Dutch city and even at the start of the twentieth century it still was the third Dutch city in terms of industrial production.

Identity in city marketing

Each city must distinguish itself from others to compete. Visitors and users need an image of the typical characteristics of a city, which draws them back to it instead of visiting a competitor. What does this mean in terms of communication? In consumer research, the product image is often divided into cognitive and affective components. The cognitive components relate to our convictions about our knowledge of the attributes of a destination, whereas the affective components involve our feelings about this. The visitor should have a favourable collection of ideas, convictions and impressions of the city. This picture, determined both cognitively and affectively, must correspond with the one that the city wants to portray. The actual image has to correspond with the desired image. Whenever the actual image differs from the desired image, as is often the case nowadays, a process has to be started during which the actual image will transform into the desired image.

What does this process entail? It means that the city will not only think about how it formulates its message and the choice of media it uses but also about the substance of its identity. It is necessary to dwell on the identity of a city, as only an image based on the identity will stick. This is pure common sense. When, on the other hand, the reality is different from people's perceptions, then this perception will gradually disappear and be replaced by the consistent characteristics of a city. So the goal of this strategy is to project those parts of a city's identity that correspond with the desired image. This can be referred to as the projected identity. This projected identity therefore consists of the parts of the identity to which the city communications department wants to draw as much attention as possible. That is why the identity is given such an important role in the strategy of change.

The identity (Pellenbarg 1991: 3) is a very brief description of what a city actually is and how it differs from other cities. The term identity is often confused with the word image. The image is the picture that certain groups have. This picture does not necessarily correspond with the identity, i.e. what a city actually is.

A city's identity is determined by just a few of its characteristics. The identifying characteristics of a city must satisfy three criteria: centrality, difference and historicity. Firstly, the city must be adequately described by just a few terms, the so-called central terms. Whoever determines the identity of a city (a researcher, the city or a group of residents) will always come up with a statement that characterizes a city on the basis of something important or essential.

The problem with the second criterion, differentiating the character of the relevant components of the identity, lies elsewhere. As differentiation is a question of relationships, a city must clarify the types of competitors it wants to distinguish itself from. The opinions of the city cannot be trusted in this context because this will lead to the so-called paradox of uniqueness: people defining as distinctive those characteristics of themselves that others consider to be rather common (van Rekom 1998: 8).

Finally, the criterion of permanence states that the characteristics and differentiating elements need to have existed for a long time. That is why the identity of a city always has a historic basis. This criterion is probably contrary to the wishes of many cities who wish to project a modern appearance.

The identity can consist of eight elements: situation, history, appearance, inner life, symbols, behaviour and communication.

The situation is a city's geographical location: natural connections (pole position), accessibility (logistic position), characteristic barriers, ties (network position) and nature (ecological position). Social geography will be able to make a lot of interesting judgements with respect to this aspect.

History is the past of a city, as described in history books. The events and persons described have to be internationally renowned. Cities revitalize certain historical events and persons and pay considerable tribute to them. The archivist can often provide a lot of information about the history of the city.

The appearance of a city gives the identity its quality. The architectural quality usually has a historical dimension: the city centre and the monuments often determine the appearance of a city. Appearance is therefore an element which is mostly studied within the discipline of architecture.

Size is the quantitative aspect of the identity. According to research by Culture Plus (2002) cities consider this element to be so important that they mention it spontaneously when interviewed about their identity. They experience it as a dichotomy between urban and rural. This element seems to be the domain of urban planners.

Inner life is the motivating elements of the identity that are often invisible. The inner life consists of the driving forces behind a city, those factors that underlie the other elements, such as pride in the city, other shared standards and values and, possibly, even a shared vision. Cultural anthropology will show the way in this area.

The symbols of a city are the elements that stand for other elements. Symbols might be seen as 'pictures, which, on the one hand, refer to characteristics of the city and, on the other, enable users to easily remember the city. Anything could be a symbol, from location (Buursink 1991) to standards (van Riel 1992), from buildings to events, from products to games.

The next element in the identity of a city is its typical behaviour. The behaviour of a city is the totality of regularly recurring and characteristic actions of the inhabitants, employees and visitors and their typical responses to external circumstances. The behaviour of the institutional municipality will be a very visible subelement within this element, partly because it determines the development of certain other elements of behaviour, such as situation and size, and partly because it is capable of strongly influencing the behaviour of its residents. The municipality often has a protagonist such as the mayor.

Communication refers to the last element of the identity, its self-presentation, i.e. the offering of pictures and messages as these are sent out.

City marketing often restricts itself to inner life, symbols and communication. In such cases, the playing field is occupied by advertising agencies, who determine how the game of city marketing is played. If cultural anthropologists want to join this game they will have to follow the rules of these masters in communication.

So, the identity of a particular city will be composed from a blend of some of the eight elements. In turn, each of these elements can consist of a number of subelements. In principle, the utility of all subelements will have to be investigated. They are only useful if they are central, different and historical.

One more point must be made. The eight elements which can determine the identity of a city are not equally stable, even though it is easy to equate identity with stability. The elements of situation and history are scarcely subject to change and are therefore called the structural elements of the identity. Size, appearance and inner life, however, do modify, even though it might take at least one generation for a change of some substance. These are therefore termed the semi-static elements of the identity. The last three elements change rather easily and cannot therefore play an important role in the identity. They are too unstable for that.

The difference in the rate of change (Noordman 2004) between the elements of the identity is shown in Table 1.

Structural	*Semi-static*	*Colouring*
History	Inner life	Symbols
Situation	Size	Communication
	Appearance	Behaviour

Table 1 The elements of identity and their rate of change.

The five structural and semi-static elements of the identity together form the personality of a city. A city's personality therefore comprises the elements that are characteristic for it, distinguish it from its competitors and finally have existed for decennia if not centuries. The identity of the tiny city of

Buren in the Netherlands, for example, is still determined by it having been a separate county centuries ago.

The other three elements give colour to the city. One has to be aware that these, being subject to change easily, rarely form an element of the identity. That only happens when they fulfil the three criteria for identity: typical, present for a long time and different from others. If these elements colour the identity of a city, then its identity is enhanced. However if, as is often the case, they play too strong a role in the marketing of a city then they will cast a shadow on the city's identity.

So the identity of a city consists of its personality and under certain conditions its colour.

Symbols for changing the image

Cities currently favour logos and slogans in their communications. This started in the 1980s. And nowadays, 81 per cent of Dutch cities use a logo (Culture Plus 2002). Also, almost every city has a slogan. Formerly, most cities used their coat of arms. Nowadays, only about 40 per cent of Dutch cities use heraldic devices. Why is this?

By introducing a logo and a slogan a city is trying to establish a modern image with outsiders. 'Sint Vite has to be eliminated from the arms of our city!' was the headline of the leading Dutch newspaper:

> Mayor and aldermen of Winschoten want a fresher logo. 'We want a more fashionable and modern image,' said Christian Democrat alderman G. v. d. Wal explaining the controversial decision of the mayor and aldermen. And St Vite, who in the fifteenth century came from the German monastery of Corvey towards the Northern part of the Netherlands, does not fit in this image.

> *NRC-Handelsblad* 28 December 2004

Cities attach considerable importance to a modern image (Culture Plus 2002). Unfortunately, all cities are communicating their endeavour to further modernity in the same manner. They already look very similar because all city centres now have the same retail chains and franchise formulas. By adding an identical means of communicating they are losing any distinctions they might have had.

As was shown earlier, symbols as well as behaviour and communication can express personality. Symbols are 'metaphors', which can characterize the personality as a whole. Behaviour can stress particular elements of the identity. Festivals are often a meaningful phenomenon. They might stress the elements of history or inner life as expressed in folklore. Finally, communication might express all elements of the personality. Moreover, symbols are able to empower and support whatever happens in the areas of behaviour and communication (van Riel 1992). Table 2 explains this.

$\rightarrow$ Symbols
Personality $\rightarrow$ Behaviour $\rightarrow$ Symbols
$\rightarrow$ Communication $\rightarrow$ Symbols

Table 2 Possible roles of symbols within the identity of a city
($\rightarrow$ means 'is empowered by').

Symbols offer more extensive possibilities for colouring a city. As well as forcefully characterizing the personality, they possess the capacity to empower behaviour and communication. This suggests that symbols should be preferred to slogans and logos. In any case, symbols are more useful than slogans and logos if a city wants to change its image, as will be shown.

Let me start with the slogan. The purpose of a slogan is to characterize an attractive behaviour of the city so clearly that it appeals to outsiders.

The main problem with slogans is that they do not care about the personality of a city. That is why slogans – such as '*Rotterdam durft*' (Rotterdam dares!) – do not tell us much. This characteristic does not differentiate Rotterdam from many other cities. Moreover, it does not refer to the city's history. Which city is not innovative and does not show vision? Because every city possesses this characteristic behaviour that is meant to show modernity, it is not relevant for the identity. Slogans should draw attention towards the personality of a city and even confirm it. Slogans failing to do so will soon loose their expressive power. We will take the city of Delft as an example of this. Ten years ago the desired image strived for by the mayor and aldermen was to become a city of knowledge workers, attractive for high tech (a 'knowledge-based city'), a place where new science would develop in such a way that it would lift the local economy. In order to realize this, Delft introduced the slogan 'Delft knowledge city'. Does Delft still have this image of being a city where one can live and work in an environment as stimulating as Cambridge (Massachusetts)? The answer is certainly not!

This slogan, which initially seems sympathetic and friendly, is not effective. In the first place it has not accomplished what it set out to do: typify the characteristic behaviour so that outsiders felt attracted. The problem was and still is that many staff of the Technical University (TU) lives outside Delft and feel emotionally connected with their home town, and not with Delft. But a more serious problem is causing trouble here. This slogan is rapidly losing its expressiveness. This is because 'knowledge city' in its full meaning does not cover the personality of Delft. It is actually – and even this only to a certain extent – a 'technology city' due to the presence of the Technical University, TNO, design bureaus, DSM yeast factories, etcetera. Moreover, the slogan is not differentiating. There are many knowledge cities in the Netherlands. Every town with a university is associated with the term knowledge city. Along the A13 motorway near Delft there is even a knowledge boulevard. TU and TNO, who are the main characterizing elements when it comes to knowledge in Delft, do not connect themselves with Delft but with partners on a provincial, national and international level. This example clearly shows that slogans will fail to gain any expressive power unless they reflect a current, clear and differentiating personality. Moreover slogans rapidly become obsolete.

Another reason for the failure of the slogan 'Delft knowledge city' is that Delft's logo does not refer to its being a knowledge city. Its logo is a cobalt blue, rippling, short line. This logo brings to mind its cute canals. Delft does indeed have a split personality: partly an old city centre and partly the TU premises. The logo actually emphasises these two separate identities of Delft. Its first identity, of course, runs counter to the desired image of a knowledge city: a schizophrenic situation.

What is the purpose of logos? A logo is often seen as a symbol in itself. It is meant to create brand value for a city. Nowadays, communication departments immediately want to develop their cities into brands. By introducing a new logo and replacing the saint in its coat of arms, Winschoten expects to add the value of modernity to its brand, whereas the value of being old-fashioned is supposed to do damage to it. One might ask if cities are on the right track with this approach. The example of Delft has shown that logos have the danger of sending out their own message or being meaningless. However, this does not have to be the case. It is possible to establish a logo with expressiveness, as Cape Town has shown with the Tafelberg (ter Berg, this volume). In this logo a landmark, i.e. a symbol, has been used.

Cities expect too much of logos if they want them to replace old symbols and acquire symbolic value themselves. Logos can be better viewed as efficient communication tools and nothing more. As such, they can take in symbols, be they landmarks or icons, Table Mountain or St Vite.

Slogans do not possess the capacity to change the image of a city, mainly because they gradually become obsolete. Logos are not able to do this either, because they are basically an opportunistic device for cities to show that they are not old fashioned.

Changing the image of a city by using symbols

Symbols, however, do possess that capacity. What would be a proper way to colour the projected personality and call attention to the city? The answer to this question is shown by the Guggenheim Museum in Bilbao. The element of appearance has changed with this piece of audacious architecture. The element of size has changed too, because at once the centre of the city was provided with a landmark. The image of this city has changed rapidly because the colouring elements of the identity were also heavily influenced. A powerful new symbol became available. The city confirmed the importance of the new museum in its behaviour and generated a huge amount of publicity for it: both news flashes and in-depth interviews. With the establishment of this landmark, Bilbao has rapidly lost its image of an industrial town, which it wanted to get rid off (Short 1999).

This example shows that landmarks possess an enormous potential to change the image of a city. However, not every city is able to construct a landmark that will put all other interesting buildings and statues in its shadow, as Bilbao did. Yet every city is certainly capable of mining its history and finding icons. Icons such as Erasmus for Rotterdam or Anne Frank for Amsterdam have the same potential as landmarks when it comes to referring to the characteristics of the city. Of course, some symbols possess more expressive power than others. Landmarks and icons have the strongest power in this regard. This is partly because these kinds of symbols find their basis in the static, invariable elements of the identity. For landmarks this is the element of situation of a city; for icons its history. A second cause, however, is certainly that symbols serve as the first point of orientation for everybody, not only mentally but also corporally. These two factors explain why landmarks and icons have such a huge symbolic power (Nas 1993).

Therefore, the best way for a city to try to realize the desired change in its image is to make use of a landmark or icon, whichever it possesses. To be feasible the symbol or icon concerned needs to be strong. It has to remain available as a point of orientation towards other locations and other historical items, whatever meanings outsiders are projecting on it. The Eiffel Tower, Big Ben and the Statue of Liberty possess the highest degree of strength, because they are unique (Nas, referred to in Sluis 2003). The first step of an applied theory of urban symbolism would probably be to look for all possible landmarks and icons of a particular city, make a proper inventory of these and then select the strongest one.

Delft will serve as an illustration of how this search could be done. We started to look for statues. Statues are possible symbols but they are not very explicit in Delft. Icons and landmarks of cities can often be found on stamps. For Delft this yields William of Orange, Hugo de Groot, Van Leeuwenhoek and Delft Blue. Furthermore, symbols were found in the subjects of paintings (View of Delft, Vermeer's Alley) and in its rituals (burials of the members of the Dutch royal family).

In Delft, the first step of this application of urban symbol theory provided four strong symbols. These are all icons, which suggest that Delft has an exceptionally interesting history. They are: Delft Blue, TU, Johannes Vermeer and William of Orange.

Which of these four symbols is the most useful for the necessary change of image? Only one symbol is needed, simply because it is confusing for the audience to be confronted with four different symbols at the same time. The second step of an applied theory of urban symbolism would be to select one of the symbols found in the first stage of the research project.

How was this done in the Delft case? The urban symbolism literature suggested looking for a hierarchy in the impact on the spectator (Nas, referred to in Sluis 2003). For city marketing purposes this even seems an obvious approach. The hierarchy is based upon the degree of fame of the symbol concerned. Discussions about fame take into account two elements: width (the countries that are familiar with the symbol concerned) and depth (the average number of inhabitants of a country that is familiar with that symbol). For the symbols of Delft a hierarchy was indeed found. Topping the list is the world famous Delft Blue. This has a certain brand value in itself. Just underneath comes the TU, world famous for its civil engineers. This is followed by the tiny picturesque canal town, known to many thanks to the paintings of Johannes Vermeer. William of Orange, who was murdered in Delft and who is important for the national history of the Dutch, occupies bottom place.

In this case the hierarchy in the impact on spectators was not decisive. Three symbols fell off due to other criteria. Off course, William of Orange is not known to most foreigners and therefore does not function as a strong symbol for them. The main problem with William of Orange as a symbol, however, is its association with death. William of Orange was killed in Delft. He was buried in the Nieuwe Kerk in Delft and since then most members of the Dutch Royal family have been buried there. It is curious to note that the author Kostof (1991) compares Delft with other cities on the basis of the Black Death, death by starvations and wars. This symbol then drops out due to its negative connotation. Not only does it mean nothing to foreigners, it leaves the Dutch, who do know the symbol very well, too little room to project their own emotions and knowledge upon the symbol. Along the way we have found a second criterion for selecting symbols. This is that the symbol has to leave enough possibilities for the spectators to associate freely.

The TU has certainly been a symbol for Delft in the past. Whoever mentioned Delft meant TU (then Technische Hogeschool) and whoever said *Delftenaar* (literally inhabitant of Delft) was thinking of the civil engineers who had completed their study at TU Delft and were building bridges and canals all over the world. However, these associations have become increasingly vague. The term TU has therefore lost its allure and will not be able to serve our purpose. This experience tells us that the symbol to be chosen should not have become vague in recent decades.

Finally, Delft Blue had to be cut out. The reason for this was that the tourists who visit Delft for Delft Blue, usually only visit the Porceleyne Fles, the last factory that still produces Delftware and which lies outside the city centre, but not the city of Delft itself. Delft Blue now has become a sort of a general slogan for the tourist industry. Delft Blue does not mean much for the inhabitants of Delft. Although Delft Blue is known throughout the world, what the term signifies has almost nothing to do with Delft. Delft Blue symbolizes The Netherlands more than Delft. Almost nothing of Delft, for instance, can be seen on the objects of Delft Blue. In Anglo-Saxon countries 'Delftware' is a rather general denominator for ceramics. Blue refers to the Dutch sky and freshness. All things considered, Delft Blue is more a 'critical period' for the identity of Delft than a symbol. This teaches us that even if a symbol is world famous, it does not necessarily refer to the city, with which it at first glance is connected.

By using additional criteria, Johannes Vermeer was eventually selected as the appropriate symbol. He has to bring about the desired image of Delft. In Delft we learned that for the selection stage, the concept of impact on the spectator, as used in the theory of urban symbolism, has to be refined. In the first place, the selected symbol must possess a positive symbolic value: because William of Orange sets the minds of the spectators on burials, he could not be used for our purpose. Moreover, the symbol chosen should not symbolize something wider than the city, as is the case with Delft Blue that actually symbolizes a whole country. So the symbol has to be effective at the right level of symbolism (Nas, in Sluis 2003). Of course the symbol should not be too vague either.

The case of Delft clearly shows that the theory of urban symbolism can contribute to city marketing by, for example, accelerating the process of changing a city's image. It will do this by successively finding all the symbols a city possesses, checking their symbolic values, finding out their poles and power, and determining on which level they operate. With this approach an applied theory of urban symbolism will be able to select the right symbol for the job.

City marketing theory

How does this first outline of an applied theory of urban symbolism relate to the existing city marketing theory? Kotler *et al.* (1999) especially favour symbols. They more or less suggest that in city marketing symbols have to replace the product – the first P in the marketing mix in conventional marketing theory. This is an interesting idea, because it solves the notorious difficulties (Noordman 2004) of defining the term product in city marketing. According to this idea, which Kotler *et al.* do not express as such, city marketing would boil down to the selling of symbols.

One gains the impression, however, that city marketeers do not have a proper grasp on symbols. Kotler *et al.*, for instance, categorize symbols as: visual symbols, verbal symbols, outstanding repeating events, and inhabitants that have become personages.

Visual symbols: logos and markers
Many authors have written about logos. We consider these to be a communication device. As such, they do not have any symbolic value and possibly even compete with symbols. However, this is not necessarily the case. Exceptions are possible, as has been shown with the Tafelberg.

Markers form an entirely different class. They are gradually gaining more attention and are starting to be considered as the most promising instrument for communicating about regions and attractions. They have been an important instrument to communicate about cities for a long time: Eiffel Tower for Paris, Big Ben for London and so on (Pilon and Huybregtse 2002).

Powerful phrasings: slogans, themes and positions
Amongst many other things, Kotler *et al.* (1999: 169) refer to slogans (Budapest: a city with a thousand face), themes (leading to the reforming of the Estonian Investment Agency) or a special position acquired (Munich: Insurance City number one in Europe). We have explained why slogans and themes have to be considered as the concisely summarized typical behaviour of a city and not as a symbol. A typical position is part of the personality and

as such not a symbol.

Outstanding events and memorable deeds
Here Kotler *et al.* mainly refer to festivals, such as the ones in Roskilde, Aix en Provence.

Inhabitants who have become personages
Now these are symbols again. Here we meet icons, as they have been dealt with in this article. One might think of Erasmus as a possible symbol for Rotterdam, because he was born there. Many cities have had famous persons living in their environment but they are often not aware of this (Figure 1). Archives could therefore be a valuable resource for city marketing.

This categorization of Kotler *et al.* already shows that city marketeers might profit from the knowledge of the urban symbolists. An applied theory of urban symbolism will be able to show which elements are actually symbols, which of these have a powerful meaning and which one should be selected.

Yet they do not need to communicate with each other on this elementary level. Take the colours blue (water) and orange (royal family) for Delft (Figure 2). A theory of urban symbolism, for example, might show the possibilities for actively using different layers in the symbols for city marketing. An interesting result of our research project in Delft was the observation that Delft actually uses colours as referential symbols. In this city, colours refer as symbols to other symbols that again are colours. 'The other colours of Delft' clearly refer to the meaning of the colours to communicate about different kinds of attractions. An applied theory of urban symbolism might provide guidance on how to distribute the colours. It might be able to give each colour its appropriate meaning, that is to say a meaning which is logically connected with the relevant static elements of the identity of Delft. At the moment the colours are chosen rather arbitrarily.

Furthermore, the boundaries of urban symbolism might become clearer. Take for instance the recent inclination of cities to build special buildings for icons with, of course, their names in huge characters on it, where the history of the icon is enlivened. In Delft one thinks of a Johannes Vermeer house (Figure 3), for which a Johannes Vermeer painting might be bought. In Rotterdam a large pressure group is active which wants to establish an Erasmus house. One would like to know if this inclination has to be recommended. Here the theory about mental mapping in urban symbolism

Figure 1 William of Orange is not known to most foreigners
(Photograph: Marijke Theunissen, 2005).

might help. Icons are meant for mental orientation. So if an Erasmus house or Vermeer centre were to be established as a visitors centre, it would enhance the possibilities of the visitors to orientate themselves mentally and corporally. However if it were to be set up as a classical museum or library it might become too grand and lose its focus.

Conclusion

Looking back at what I have written, I think that applied urban symbolism as a separate sub-discipline of the theory of urban symbolism might be feasible. It would mean using the body of knowledge developed about urban symbolism in the area of social anthropology for city marketing purposes. It would rely on a common theory for finding useful symbols, selecting symbols and loading symbols with meaning about the personality of the city.

Figure 2 The use of colours as referential symbols: A blue *velo taxi* with underneath the other colours in a row (*Ontdek de andere kleuren van Delft*) (Photograph: Marijke Theunissen, 2005).

Applied urban symbolism would most probably start with studies of best practices, for instance categorized according to the scheme of Short (Short 1999; see also Sluis 2003) to enable comparison. An adaptation of Shorts scheme would probably have to be made, because Short does not seem to be familiar with the phenomenon of historical cities, which are common in Europe.

Applied urban symbolism would eventually advise municipalities with respect to a key aspect of their city marketing policy: symbols and their use. At present, the most important selling point of an urban symbolist is that symbols can be used for making fast moves if the image of a particular city has to be changed.

This is, of course, a preliminary conclusion. Some rather important problems have not been touched upon. The first one is the necessary change in the attitude of the city marketers. They will have to be prepared to find adequate meanings and make these available. For example, if a town wants

Figure 3 'One thinks of a Johannes Vermeer house'
(Photograph: Marijke Theunissen, 2005).

more cultural tourists instead of mass tourists and its icon is Van Gogh, it will have to stop communicating the superficial meaning (i.e. van Gogh having cut off his ear that is supposed to be attractive for everybody). Instead, it will have to communicate the meaning that is attractive for cultural tourists (i.e. Van Gogh being the foremost impressionist). The city marketer will have to learn to load the chosen symbol with meaning for the target group. Secondly, the applied urban symbolists will have to develop techniques to find original elements in the identity of a city that are different from those of other cities.

References

Buursink, J. (1991) *Cities in the Market: The Elan of City Marketing*. Muiderberg: Coutinho.

Culture Plus (2002) *Stedelijk Image*. Arnhem. [Rapport uitgebracht aan de Stichting Cultuurfonds van de Bank Nederlandse Gemeenten.]

Kostof, S. (1991) *The City Shaped: Urban Patterns and Meanings through History*. London: Thames and Hudson.

Kotler, P., C. Asplund, I. Rein and D.H. Halder (1999) *Marketing Places in Europe*. London: Pearson Education Limited.

Nas, P.J.M. (ed.) (1993) *Urban Symbolism*. Leiden: E.J. Brill.

Noordman, Th.B.J. (2004) *Cultuur in de Citymarketing*. Den Haag: Elsevier Business Publishers.

Pellenbarg, P.H. (1991) *Identity, Image and Economische Ontwikkeling van Regio's*. Groningen: Geo Pers.

Pilon, A. and M. Huybregtse (2002) 'Blikvangers positioneren en imagecampagnes moeten beeld van Nederland opfrissen.' *Recreatie en Toerisme*, no. 2, pp. 32–3, 35.

Rekom, J. van (1998) *Corporate Identity: Development of the Concept and a Measurement Instrument*. Rotterdam. [PhD thesis, Erasmus University.]

Riel, C.B.M. van (1992) *Identity and Image: Een Inleiding in the Corporate Communication*. Schoonhoven: Academic Service.

Risbo (2004) *De Identiteit van Delft*. Rotterdam.

Short, J.R. (1999) *The Urban Order: An Introduction to Cities, Culture and Power*. Cambridge: Blackwell Publishers Inc.

Sluis, R.J. (2003) 'A reassessment of Lynch's Conception of Meaning.' In: P.J.M. Nas, G. Persoon and R. Jaffe (eds.), *Framing Indonesian Realities: Essays in Symbolic Anthropology in Honour of Reimar Schefold*, pp. 251–73. Leiden: KITLV Press.